Nano-Biotechnology for Biomedical and Diagnostic Research

Advances in Experimental Medicine and Biology

For further volumes:
http://www.springer.com/series/5584

Eran Zahavy • Arie Ordentlich • Shmuel Yitzhaki
Avigdor Shafferman
Editors

Nano-Biotechnology for Biomedical and Diagnostic Research

Springer

Editors
Eran Zahavy
Israel Institute for Biological Research
P.O. Box 19
Ness-Ziona, 74100
Israel
Eranz@iibr.gov.il

Arie Ordentlich
Israel Institute for Biological Research
P.O. Box 19
Ness-Ziona, 74100
Israel
Arieo@iibr.gov.il

Shmuel Yitzhaki
Israel Institute for Biological Research
P.O. Box 19
Ness-Ziona, 74100
Israel
Shmuely@iibr.gov.il

Avigdor Shafferman
Israel Institute for Biological Research
P.O. Box 19
Ness-Ziona, 74100
Israel
oholo@iibr.gov.il

ISSN 0065-2598
ISBN 978-94-007-2554-6 e-ISBN 978-94-007-2555-3
DOI 10.1007/978-94-007-2555-3
Springer Dordrecht Heidelberg London New York

Library of Congress Control Number: 2011943355

Printed on acid-free paper

Springer is part of Springer Science+Business Media (www.springer.com)

Preface

The new volume in the Advances in Experimental Medicine and Biology entitled: "*Nano Biotechnology for Biomedical and Diagnostics Research*" will address the novel application of nanotechnology and nano-materials in biological research, diagnostics and therapeutics. Specifically the book will address research aspects related to nanomaterial in imaging and biological research, nano materials as a bio-sensing tool, oligonucleotides and biopolymer as nano building blocks, nano materials for drug delivery, medicinal and therapeutic application and cyto-toxicity of nano-materials. All of the chapters in this book were presented during the ***47th Oholo conference*** entitled: ***Novel Approaches in Nano Bio-Technology for Biomedical and Diagnostics Research***, which took place in Eilat, Israel in December 2010.

The manuscripts in the book intend to present specifically biological, diagnostic and medical problems with their potential solution by nano technology or materials. For example in the diagnostics and imaging field the will discuss the contribution of q-dots as imaging tools, fluorescent labels, their application as donor in FRET pair and more. Gold nanoparticles will be presented as SERS and electrochemical enhancers.

In the therapeutic field the book will present the potential of using smart nanoparticles in air liposomes with the combination of acoustic wave in various possible treatments.

The subject of forming functional nano structures from biological building block, mainly oligonucleotides will present. This will be shown as building block for molecular machines, functional networks or unique origami structures.

In these subjects this volume is expanding the possible application of nanotechnology in the major field of medicine and biology. In this respect this book is unique, since it would arise from the biological problems to the nano technology possible solution and not vice versa.

Contents

Chapter 1
Biomolecule/Nanomaterial Hybrid Systems for Nanobiotechnology

Ran Tel-Vered, Omer Yehezkeli, and Itamar Willner

Abstract The integration of biomolecules with metallic or semiconductor nanoparticles or carbon nanotubes yields new hybrid nanostructures of unique features that combine the properties of the biomolecules and of the nano-elements. These unique features of the hybrid biomolecule/nanoparticle systems provide the basis for the rapid development of the area of nanobiotechnology. Recent advances in the implementation of hybrid materials consisting of biomolecules and metallic nanoparticles or semiconductor quantum dots will be discussed. The following topics will be exemplified: (i) The electrical wiring of redox enzymes with electrodes by means of metallic nanoparticles or carbon nanotubes, and the application of the modified electrodes as amperometric biosensors or for the construction of biofuel cells. (ii) The biocatalytic growth of metallic nanoparticles as a means to construct optical or electrical sensors. (iii) The functionalization of semiconductor quantum dots with biomolecules and the application of the hybrid nanostructures for developing different optical sensors, including intracellular sensor systems. (iv) The use of biomolecule-metallic nanoparticle nanostructures as templates for growing metallic nanowires, and the construction of fuel-driven nanotransporters.

Keywords Quantum dots • Bioelectronics • Sensor • Nanoparticles • Nanowires • Nanotechnology

Nanomaterials, such as metallic nanoparticles (NPs), semiconductor quantum dots (QDs) or carbon nanotubes (CNTs) exhibit unique electronic (Katz and Willner 2004; Shipway et al. 2000), catalytic (Alivisatos 1996), and optical (Kelly et al. 2003) properties. Furthermore, biomolecules, such as enzymes, antigen/antibodies or DNA, exhibit similar nano-dimensions and reveal unique catalytic and recognition properties. Thus, the integration of the biomolecules with the nanomaterials may lead to hybrid nanostructures that combine the properties of the two components and yield materials of new functionalities. Furthermore, nanotechnology provides new microscopy tools to image and manipulate surfaces, and to characterize chemical interactions at the molecular level. All these paved the way to advance the rapidly developing research of bionanotechnology (Niemeyer 2001). The present article aims to introduce the potential applications of nanobiotechnology by discussing and highlighting some topics that have been developed in our laboratory.

R. Tel-Vered • O. Yehezkeli • I. Willner (✉)
The Institute of Chemistry, The Hebrew University of Jerusalem, Jerusalem 91904, Israel
e-mail: willnea@vms.huji.ac.il

E. Zahavy et al. (eds.), *Nano-Biotechnology for Biomedical and Diagnostic Research*, Advances in Experimental Medicine and Biology 733, DOI 10.1007/978-94-007-2555-3_1,

1.1 Electrical Contacting of Enzymes with Electrodes for the Construction of Amperometric Biosensors and Biofuel Cells

The electrical contacting of redox enzymes with their macroscopic environments, e.g., electrodes, is one of the most fundamental topics in bioelectrochemistry (Willner and Willner 2001; Willner et al. 2006a; Heller 1992; Wang 2008). While usually redox enzymes do not electrically communicate with electrodes, electron transfer cascades dominate important enzymatic processes in nature. The lack of electrical communication between the redox centers of proteins and electrodes may be rationalized in terms of the Marcus electron transfer theory, Eq. 1.1, stating that the electron transfer rate between a donor and acceptor pair is controlled by the distance, d, separating the units, the electronic coupling constant, β, the free energy associated with the electron transfer process, ΔG°, and the reorganization energy accompanying the electron transfer process, λ (Marcus and Sutin 1985).

$$k_{ET}\,\alpha\, exp\left[-\beta\left(d-d_o\right)\right]\cdot exp[-(\Delta G^0+\lambda)^2/\left(4RT\lambda\right)] \tag{1.1}$$

The redox center in the protein and the electrode support may be considered as an electron donor-acceptor pair. Since the redox center is usually embedded in the protein, a spatial separation of the active components in the electron transfer process typically occurs, leading to low electron transfer rates, according to Eq. 1.1. Fortunately, Mother Nature has resolved the barriers for electron transfer by orienting the redox-proteins into complexes exhibiting intimate distances between the redox centers, or through the participation of diffusional redox-active cofactors that mediate the electron transfer between proteins. Different man-made approaches to mimic nature and to electrically contact redox enzymes with electrodes by shortening the electron transfer distances and the structural alignment of the redox enzymes in respect to the electrodes were developed. These included the application of diffusional electron transfer mediators (Bartlett et al. 1991), the chemical modification of the proteins with electron relays (Schuhmann et al. 1991; Willner et al. 1994), the reconstitution of apo-proteins on relay-cofactor units assembled on electrodes (Willner et al. 1994; Zayats et al. 2002), and the immobilization of the proteins in redox polymers (Maidan and Heller 1992; Rajagopalan et al. 1996). The development of the different means to establish the desired electrical communication enabled the development of different amperometric biosensors and biofuel cell elements. Mechanistic studies using the different means revealed, however, only limited electron transfer exchange rates between the enzymes and the electrodes. A fast electron transfer between the redox centers and the electrodes is essential to construct sensitive, selective and miniaturized amperometric biosensors (Murphy 2006), and to yield high power output biofuel cells (Barton et al. 2004; Heller 2004; Willner et al. 2009). The availability of conductive nanomaterials, such as metal nanoparticles (NPs) or carbon nanotubes (CNTs) introduced new opportunities to control the electrical contacting between the redox enzymes and the electrodes by the engineering of enzyme/nanomaterial hybrid nanostructures on electrode surfaces (Willner et al. 1996; Xiao et al. 2003). Figure 1.1A depicts the construction of an electrically contacted Au NP/glucose oxidase (GOx) hybrid monolayer on an electrode surface. Benzene dithiol (**1.1**) was assembled on a Au electrode surface and Au NPs (1.4 nm) functionalized with a single N-hydroxysuccinimide active ester were linked to the monolayer. An aminoethyl flavin adenine dinucleotide, (**1.2**), amino-FAD, was covalently linked to the Au NPs, and apo-glucose oxidase (apo-GOx), the enzyme lacking its native FAD cofactor was reconstituted onto the cofactor sites linked to the electrode. The reconstitution process led to the optimal alignment of the enzyme redox center in respect to the electrode, and positioned the Au NPs, acting as relay units, between the redox center of the enzyme and the electrode. Indeed, the resulting enzyme electrode revealed an effective electrical contact between the enzyme and the electrode, and upon the application of the required potential the bioelectrocatalyzed oxidation of glucose was activated. The resulting amperometric responses were controlled by the concentration of glucose, Fig. 1.1B. Knowing the surface coverage of GOx on the

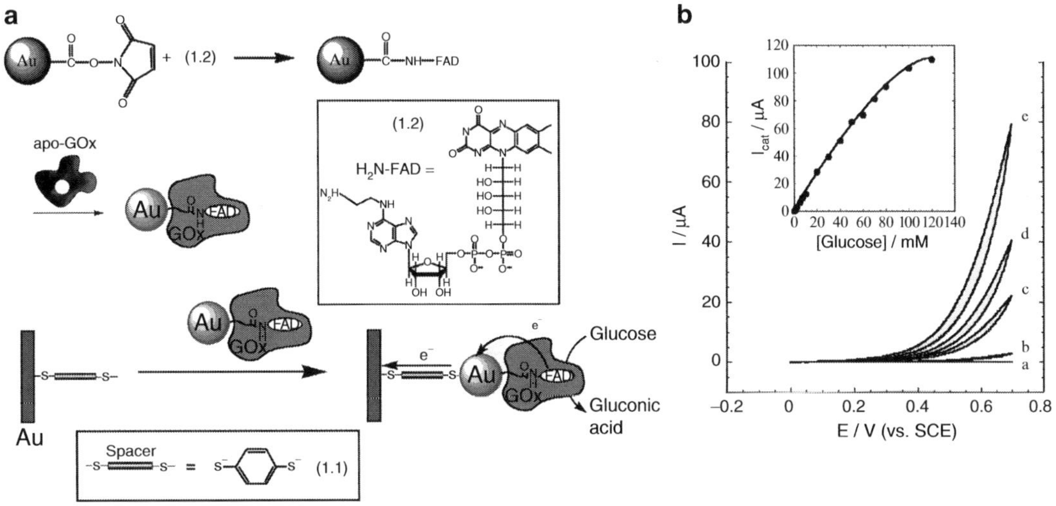

Fig. 1.1 (**a**) The assembly of a Au NP (1.4 nm) electrically-contacted glucose oxidase electrode by the reconstitution of apo-GOx on an FAD-functionalized Au NP, and the immobilization of the enzyme/nanoparticle hybrid on an electrode surface using a benzene dithiol crosslinker. (**b**) Cyclic voltammograms corresponding to the bioelectrocatalyzed oxidation of variable concentrations of glucose by the electrically-contacted Au NP-reconstituted GOx-modified electrode. Glucose concentrations correspond to: (*a*) 0 mM, (*b*) 1 mM, (*c*) 2 mM, (*d*) 5 mM, and (*e*) 10 mM. Inset: Calibration curve corresponding to the electrocatalytic currents at different glucose concentrations. *Part B* – Reproduced from Xiao et al. (2003), reprinted with permission from AAAS

electrode, and from the saturation current generated by the enzyme electrode, the turnover rate of electrons between the redox center and the electrode was estimated to be 5,000 $e^{-}\cdot s^{-1}$, a value that is ca. eight-fold higher than the electron transfer turnover rate between the enzyme redox site and its native acceptor, O_2. This effective electrical contact between the redox center and the electrode was attributed to the structural alignment of the enzyme redox center in an optimal steric position in respect to the electrode, and to the precise positioning of the Au NPs acting as electron relays in between the redox center and the electrode. The effective electrical contacting led to a sensitive and selective enzyme electrode for the detection of glucose that was also oxygen-insensitive and unperturbed by common glucose sensing interferants such as ascorbic acid or uric acid.

A related approach was applied to electrically contact the enzyme glucose oxidase, GOx, by means of carbon nanotubes (CNTs) (Patolsky et al. 2004b), Fig. 1.2A. Long CNTs were oxidatively cleaved to yield carboxylic acid functionalities at their ends. The resulting CNT were separated to yield length controlled fractions with dimensions 25–30 nm, 60–70 nm, 100–130 nm, and 190–220 nm. The enzyme electrodes were prepared by the primary covalent attachment of the CNTs to a cysteamine monolayer-functionalized Au electrode. The aminoethyl-FAD cofactor was attached to the vertically aligned CNTs, and apo-GOx was reconstituted on the cofactor sites to yield the electrode assembly. Accordingly, different electrodes, consisting of CNTs of variable lengths were prepared. Figure 1.2B depicts the microscopic characterization of the functionalized electrode and the modified CNTs. Figure 1.2B, Image (I), shows the vertically aligned CNTs after the reconstitution of the apo-enzyme on the surface. Evidently, after reconstitution, nano-objects with dimensions of single protein units were observed. Also, the reconstitution of the enzyme units at the ends of the CNTs was nicely confirmed by performing the reconstitution process in solution on CNTs functionalized at their ends with the FAD cofactor. Figure 1.2B, images (II) and (III), show, respectively, the AFM and TEM (stained) images of the CNTs reconstituted at their ends with the GOx. The protein units are clearly visible, confirming the reconstitution process. The resulting CNT/enzyme-functionalized electrodes exhibited electrical contact between the enzyme redox centers and the surface, and the bioelectrocatalyzed

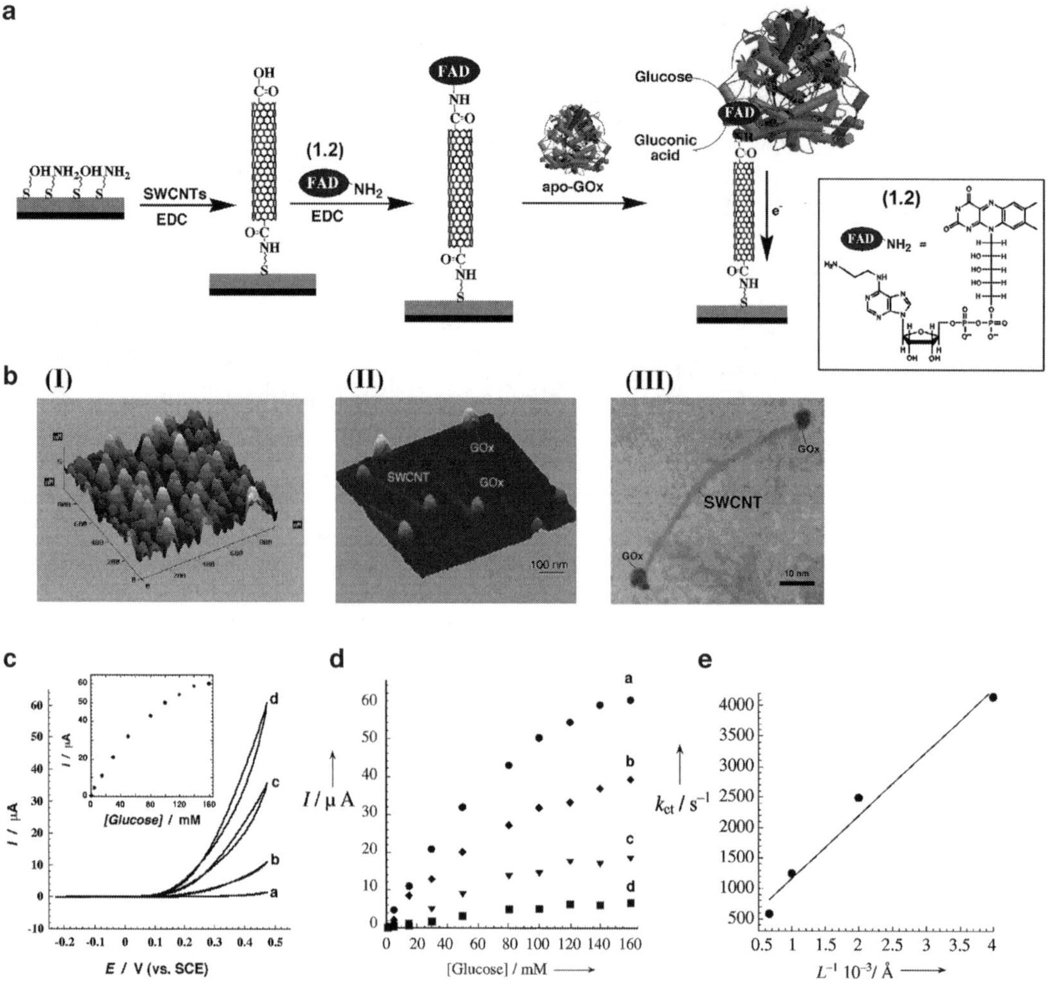

Fig. 1.2 (**a**) Assembly of the carbon nanotubes (CNTs) electrically contacted glucose oxidase electrode. (**b**) I – AFM image of the GOx reconstituted on the FAD-functionalized CNTs monolayer associated with the Au surface. II – AFM image of CNTs reconstituted at their ends with GOx units. III – HRTEM image of a CNT modified at its ends with GOx units. (**c**) Cyclic voltammograms corresponding to the electrocatalyzed oxidation of different concentrations of glucose by GOx reconstituted on the 25 nm-long FAD-functionalized CNTs assembly. Glucose concentrations: (*a*) 0; (*b*) 20; (*c*) 60; (*d*) 160 mM. Data recorded in phosphate buffer, 0.1 M, pH = 7.4. Scan rate 5 $mV s^{-1}$. Inset: Calibration curve corresponding to the amperometric responses (at E = 0.45 V) in the presence of different concentrations of glucose. (**d**) Calibration curves corresponding to the amperometric responses (at E = 0.45 V) of reconstituted GOx–CNTs electrodes in the presence of variable concentrations of glucose and different CNT lengths as electrical connector units: (*a*) ~25 nm CNTs; (*b*) ~50 nm CNTs; (*c*) ~100 nm CNTs; (*d*) ~150 nm CNTs. (**e**) Dependence of the electron-transfer turnover rate between the GOx redox center and the electrode on the lengths of the CNTs comprising the enzyme electrodes (Reproduced with permission from Patolsky et al. 2004. Copyright Wiley-VCH Verlag GmbH & Co. KGaA)

oxidation of glucose was stimulated. Figure 1.2C shows the cyclic voltammograms corresponding to the bioelectrocatalyzed oxidation of variable concentrations of glucose by the modified electrodes and the resulting calibration curve. The amperometric responses increased as the concentration of glucose was elevated, and they leveled off to a value reflecting the saturation of the active sites of the enzyme units by the substrate (and the optimal bioelectrocatalytic performance of the electrode). Knowing the surface coverage of the enzyme and the saturation current, a turnover rate of transferring electrons from the active site, through the CNTs, to the electrode was estimated to be 4,100 e^{-} s^{-1}. It was found, however, that although CNTs exhibit ballistic conductivity, the electrical contacting efficiency of the enzyme with the electrode was controlled by the lengths of the connecting CNTs, Fig. 1.2D. As the length of the connecting CNTs was increased, the electrical contacting efficiency revealed lower values. Although record distances for electron transfer in the system (~200 nm) were demonstrated, the dependence of the electrical contacting on the length of the CNTs needs an explanation. We refer to the fact that in order to assemble the enzyme/CNTs nanostructures on the electrodes, we oxidatively shortened the CNTs to introduce carboxylic acid functionalities that are essential for the construction of the systems. This process is known to introduce oxidative defects into the side walls of the CNTs (alcohol, aldehyde or carboxylic acid sites), which perturb the transport of the electrons through the CNTs, and as a result, the tunneling of electrons across the defect sites, or the back scattering of the electrons from the defect sites to a conductive conjugated domain are needed. These processes perturb the conductivity along the CNTs. Since the probability for the occurrence of defects relates directly to the length of the tubes, the electron transfer rate along the CNTs should be inversely proportional to their length. Figure 1.2E depicts a linear relation observed between the charge transport rate along the tubes and 1/l (where l is the length of the CNTs), supporting this explanation. These latter systems exemplify a central challenge in the area of nanobiotechnology, in which new, unpredictable, properties are often observed for nanomaterial hybrids at the nanoscale. In order to harness these hybrid structures for practical applications, it is important to understand the structure-function relationships associated with these nanostructures.

1.2 Catalytic Properties of Metallic Nanoparticles (NPs) and Their Implementation for Sensing and Nanocircuitry

The catalytic reduction of metal ions and their deposition on metallic NPs seeds is a well established method to enlarge metallic nanoparticles or to generate core/shell biometallic NPs. The catalytic growth of Au NPs was used for the electrical detection of DNA (Park et al. 2002), Fig. 1.3A. A capturing nucleic acid (**2.1**) that is complementary to a domain of the target DNA, (**2.2**), was immobilized on a gap surface that separated two electrodes. The hybridization of the target DNA with the capturing unit was followed by the hybridization of a nucleic acid, (**2.3**),-functionalized Au NPs that is complementary to the single-stranded end of the target DNA. The catalytic hydroquinone-stimulated reduction of Ag^{+} ions resulted, then, in the deposition of Ag° on the Au NPs and the formation of core-shell Au/Ag NPs. The intimate contact between the metallic NPs resulted in conducting paths through the gap separating the electrodes, as reflected by a decrease in the resistance of the gap upon the growth of the NPs, Fig. 1.3B, curve (a). The resistance across the gap was not controlled only by the time of the growth of the NPs, but also by the coverage of the Au NPs catalytic seeds associated with the gap, that was, in turn, controlled by the bulk concentration of the target DNA, (**2.2**), and thus, the target coverage on the gap domain. Thus, the resistivity of the gap domain at a fixed growth time of the NPs provides a quantitative measure for the concentration of the analyte. The method proved to be successful in detecting single-base mismatches in DNA. For example, exchange of the A-base in the capture nucleic acid with a G-base enabled the temperature-controlled elimination of the hybridization of (**2.2**) due to a lower melting temperature. As a result, the insulating features of the gap were preserved, Fig. 1.3B, curve (b).

The enzyme-catalyzed growth of metallic NPs, e.g., Au or Ag NPs, was used to develop optical sensors that monitor the concentrations of the substrates of the biocatalysts, or the enzyme activities, by following the localized plasmon absorbance of the resulting NPs (Willner et al. 2006b). Many enzymes generate as products reducing agents that can directly reduce metal ions to metallic NPs in the presence of the NPs seeds. For example, NAD^{+}-dependent enzymes oxidize numerous substrates while generating the reduced 1, 4-dihydronicotin-amide adenine dinucleotide cofactor, NADH. This

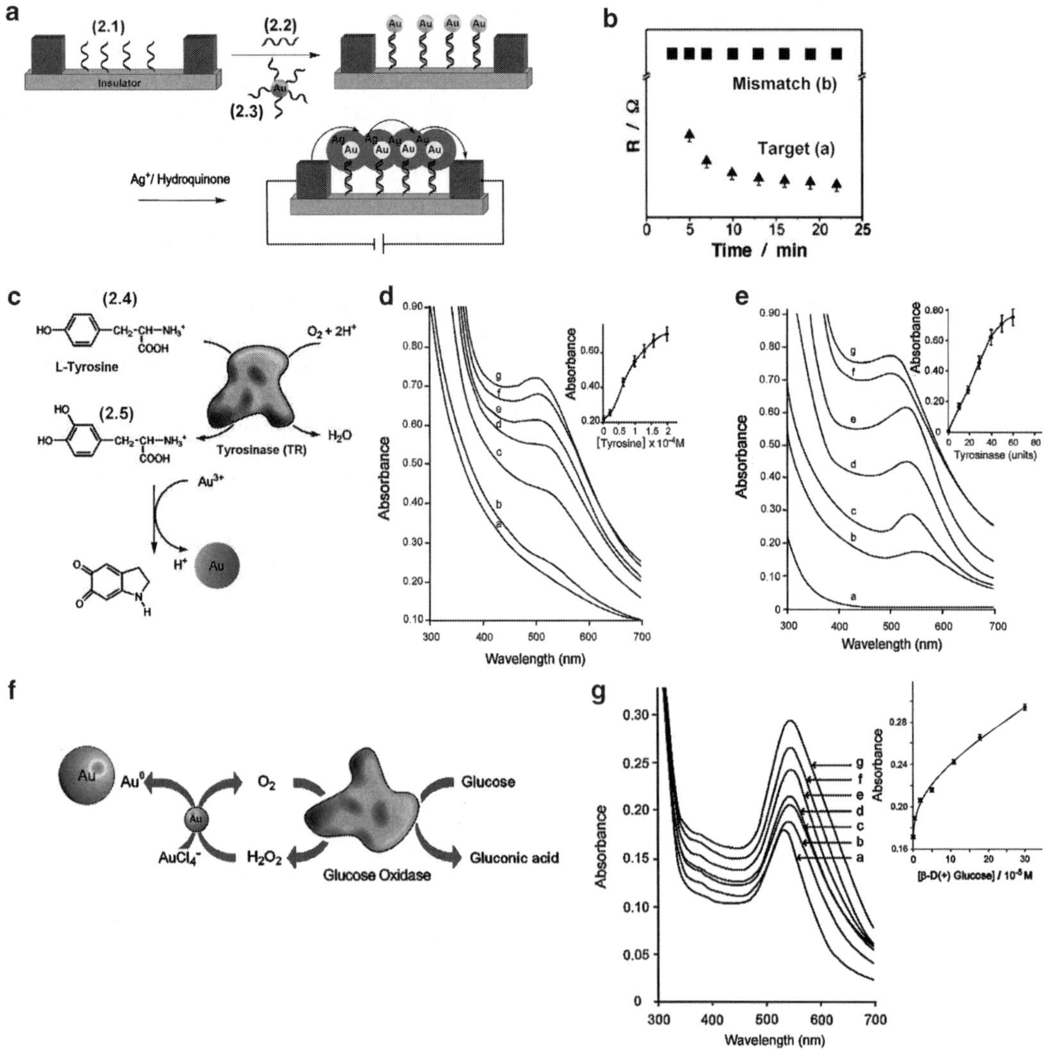

Fig. 1.3 (**a**) Detection of DNA through the generation of a Au NP-labeled nucleic acid/DNA recognition complex in an insulating gap separating two microelectrodes. The catalytic growth of the Au NPs yields conductivity paths between the microelectrodes and a decrease in the resistivity of the gap region. (**b**) Time-dependent resistance changes upon analyzing: (*a*) The target DNA, and (*b*) A single-base mismatched DNA, by the catalytic enlargement of the Au NPs labels. (**c**) Assay of tyrosinase activity through the biocatalyzed oxidation of tyrosine and the L-DOPA-mediated formation of Au NPs. (**d**) Absorbance spectra corresponding to the tyrosinase-stimulated growth of Au NPs in the presence of different concentrations of tyrosine: (*a*) 0, (*b*) 2×10^{-5}, (*c*) 6×10^{-5}, (*d*) 1×10^{-4}, (*e*) 1.3×10^{-4}, (*f*) 1.6×10^{-4}, and (*g*) 2×10^{-4} M. In all systems the reaction mixture includes $HAuCl_4$, 2×10^{-4} M, tyrosinase, 40 U mL^{-1}, and CTAC, 2×10^{-3} M, in 0.01 M phosphate buffer (PB). Spectra were recorded after a fixed time interval of 10 min. Inset: Calibration curve corresponding to the absorbance, at $\lambda = 520$ nm, of the Au NPs formed in the presence of variable concentrations of tyrosine. (**e**) Absorbance spectra of Au NPs formed by variable concentrations of tyrosinase: (*a*) 0, (*b*) 10, (*c*) 20, (*d*) 30, (*e*) 35, (*f*) 40, and (*g*) 60 U mL^{-1}. All systems include tyrosine, 2×10^{-4} M, $HAuCl_4$, 2×10^{-4} M, and CTAC, 2×10^{-3} M, in 0.01 M PB. Spectra were recorded after a fixed time interval of 10 min. Inset: Calibration curve corresponding to the absorbance, at $\lambda = 520$ nm, of the Au NPs formed in the presence of variable concentrations of tyrosinase. (**f**) The biocatalytic enlargement of Au NPs by the GOx-mediated system. (**g**) Absorbance spectra of the Au NPs-functionalized glass supports upon reaction with 2×10^{-4} M $HAuCl_4$ and 47 μg mL^{-1} GOx in 0.01 M PB that includes CTAC, 2×10^{-3} M, and different concentrations of β-D(+) glucose: (*a*) 0; (*b*) 5×10^{-6}; (*c*) 2×10^{-5}; (*d*) 5×10^{-5}; (*e*) 1.1×10^{-4}; (*f*) 1.8×10^{-4}; (*g*) 3.0×10^{-4} M. For all experiments the reaction time interval was 10 min, 30°C. Inset: The respective calibration curve, at $\lambda = 542$ nm. *Parts C to E* – Reprinted with permission from Baron et al. (2005). Copyright (2005) American Chemical Society. *Part G* – Reprinted with permission from Zayats et al. (2005). Copyright (2005) American Chemical Society

cofactor was found to reduce Au^{3+} ions and enlarge Au NPs seeds. The plasmon absorbance of the resulting NPs was used to detect the respective substrate of the NAD^+-dependent enzyme, e.g., ethanol in the case of alcohol dehydrogenase (Xiao et al. 2004). This will be exemplified here with two biocatalytic systems that lead to the growth of metal NPs and enable the optical detection of the enzyme activity or the substrate. Tyrosinase catalyzes the oxidation of phenol derivatives by oxygen to O-dihydroquinone compounds. Elevated amounts of tyrosinase are found in melanoma cancer cells, and the enzyme is considered as a biomarker for such cells. Accordingly, L-tyrosine, (**2.4**), was subjected to the tyrosinase-stimulated oxidation to L-DOPA, (**2.5**). The product reduced $AuCl_4^-$ to Au NPs, and their absorbance features provided a quantitative measure for the tyrosinase (Baron et al. 2005), Fig. 1.3C. The time-dependent increase in the plasmon absorbance of the Au NPs, at a constant concentration of tyrosinase, is shown in Fig. 1.3D, whereas the absorbance features of the Au NPs upon analyzing different concentrations of the biocatalyst, and allowing the catalytic growth of the NPs for a fixed time-interval, are shown in Fig. 1.3E. Evidently, as the concentration of the Au NPs increased, the absorbance of the resulting NPs was intensified.

A further system demonstrating the biocatalytic growth of Au NPs is shown in Fig. 1.3F, where the glucose oxidase-mediated oxidation of glucose by O_2 yielded H_2O_2, and the resulting product reduced $AuCl_4^-$ to Au^0 on Au NPs catalytic seeds (Zayats et al. 2005). Thus, the oxidation of glucose led to the growth of the Au NPs, and their absorbance features were controlled by the concentration of the substrate, Fig. 1.3G. The resulting calibration curve is shown in Fig. 1.3G, inset.

The biocatalytic growth of metallic NPs has significant implications in nanobiotechnology. The enzymatic growth of metallic NPs, i.e., Au NPs, paves the way to develop new optical sensors. The catalytic growth of metallic NPs in cellular environments does not only provide a new method to synthesize metallic NPs, but also to image intracellular metabolic processes using microscopy methods (e.g. reflectance spectroscopy). Finally, the biocatalytic growth of metallic nanoparticles holds a great promise for the bottom-up synthesis of nano-objects such as metallic nanowires for nanoelectronic devices (see Sect. 1.4).

1.3 Bioanalytical Applications of Hybrid Semiconductor-Protein Systems

Semiconductor nanoparticles (quantum dots, QDs) exhibit unique optical and photophysical properties (Alivisatos 1996). These are reflected by size-controlled luminescence spectra (Nirmal and Brus 1999), high luminescence quantum yields, narrow luminescence spectra, large Stokes shifts, and stability against photobleaching. Not surprisingly, QDs gain substantial interest as nanomaterials for bioanalytical applications (Alivisatos 2003; Gill et al. 2008; Medintz et al. 2005). Indeed, numerous studies have implemented QDs as optical labels for imaging biorecognition events, and the size-controlled luminescence properties of the QDs were implemented to design luminescence labels for parallel multiplexed analysis (Goldman et al. 2004; Gerion et al. 2003). The application of QDs for bioanalysis requires, however, the use of QDs in aqueous media, and their modification with biomolecules or molecular substrates, while retaining their unique photophysical properties (Gerion et al. 2001; Willard et al. 2001; Mattoussi et al. 2000; Medintz et al. 2006). Indeed, tremendous advances were accomplished in the past few years in the synthesis of water-soluble, chemically-modified QDs and biomolecule-functionalized QDs that retain their photophysical properties. These hybrid structures found diverse bioanalytical applications. Particularly interesting are the uses of QDs as optical labels for following the dynamics of biocatalytic processes, and for the imaging of intracellular metabolic processes. To this end, photophysical mechanisms, such as electron transfer (ET) quenching of the luminescence of the QDs, or fluorescence resonance energy transfer (FRET) were implemented to characterize the dynamics of biocatalytic processes. These possibilities will be addressed here with three different examples.

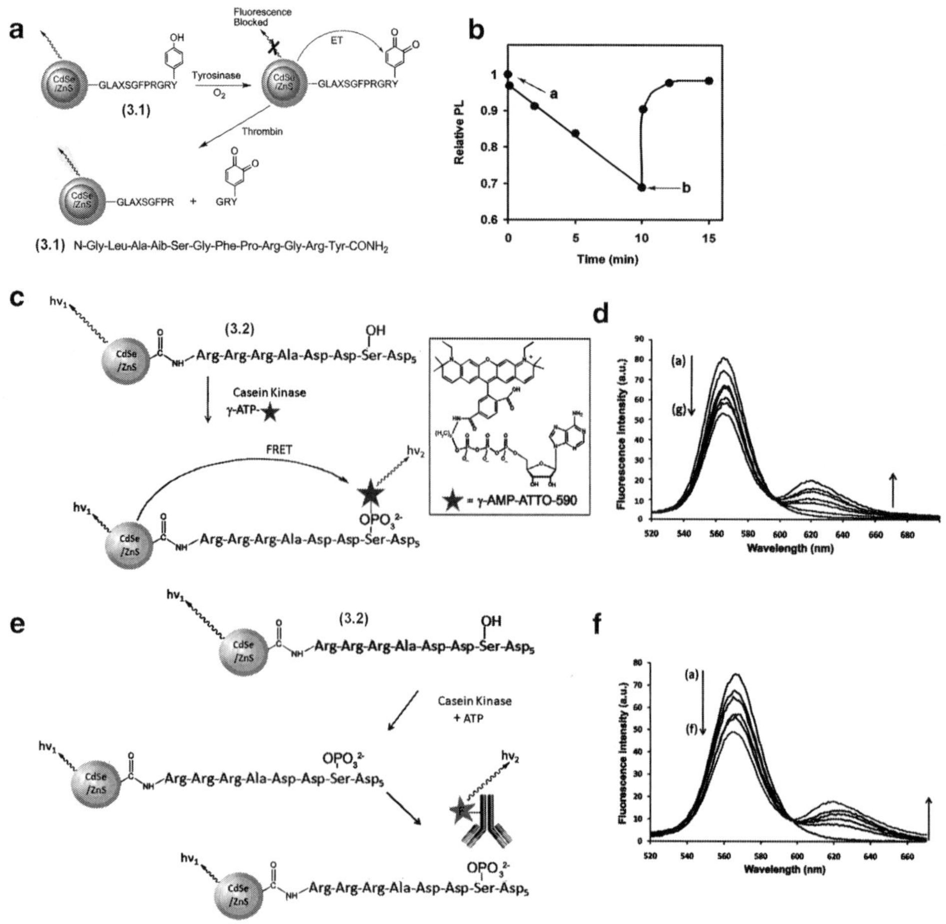

Fig. 1.4 (**a**) Probing tyrosinase and thrombin activities through the quenching of the luminescence of the QDs by the generation of the quinone-containing peptide and the regeneration of the luminescence of the QDs by the thrombin-induced scission of the peptide. (**b**) Time-dependent luminescence intensities upon reacting the (**3.1**)-functionalized QDs with tyrosinase (*point a*) and thrombin (*point b*). All data were recorded in 10 mM phosphate buffer (PB) solution, pH=6.3, under air, at 25°C. (**c**) Optical analysis of casein kinase, CK2, via FRET, by the biocatalytic phosphorylation of the (**3.2**)-functionalized QDs with γ-ATP-Atto-590. (**d**) Time-dependent luminescence spectra corresponding to the (**3.2**)-modified QDs upon treatment with γ-ATP-Atto-590: (a) Before the addition of CK2, and (*b–g*) After the interaction with CK2, 1 unit. Spectra were recorded at time intervals of 7 min. (**e**) Optical analysis of CK2 via FRET by the binding of Atto-590-modified antiphosphoserine-antibody to the phosphorylated functionalized QDs. (**f**) Time-dependent luminescence spectra of the (**3.2**)-modified QDs upon the CK2 phosphorylation of the peptide by ATP in the presence of the Atto-590-modified-antiphosphoserine-anti-body: (*a*) Before the addition of CK2, (*b–f*) After the interaction with CK2, 1 unit. Spectra were recorded at time intervals of 8 min. *Parts A and B* – Reprinted with permission from Gill et al. (2006). Copyright (2006) American Chemical Society. *Parts C to F* – Reprinted with permission from Freeman et al. (2010). Copyright (2010) American Chemical Society

CdSe/ZnS QDs were modified with the tyrosine-containing peptide (**3.1**), capping monolayer, Fig. 1.4A. The chemically-modified QDs were interacted with tyrosinase and O_2. Tyrosinase, is known to be a biomarker for melanoma cancer cells, and it biocatalyzes the oxidation of tyrosine residues to dihydroquine, and subsequently to o-quinone. Thus, the tyrosinase-mediated oxidation of the peptide (**3.1**) generated the dopaquinone residues, and these acted as electron transfer quenchers for the luminescence of the QDs (Gill et al. 2006). Figure 1.4B depicts the time-dependent luminescence of the QDs. As the oxidation proceeds, the content of the quencher units increases and the quenching process

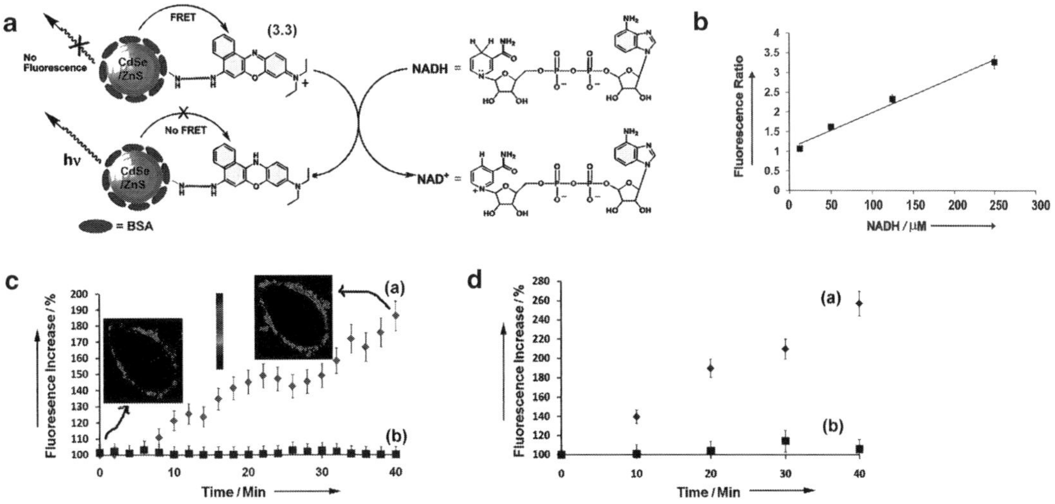

Fig. 1.5 (**a**) Sensing of NADH by Nile-blue-functionalized CdSe/ZnS QDs. (**b**) Fluorescence intensity ratios of the modified QDs after the addition of different concentrations of NADH and the fluorescence intensities prior to the addition of NADH. The fluorescence intensities following the addition of NADH were recorded after a constant time interval of 15 min. The samples included 50 nM QDs in a 10 mM PB (pH=8.8). (**c**) (*a*) Time-dependent fluorescence changes of HeLa cells that include the functionalized QDs upon interaction with 50 mM D-glucose. (*b*) Time-dependent fluorescence changes of HeLa cells that include the functionalized QDs upon interaction with 50 mM L-glucose. Each data point corresponds to the analysis of 20 different cells. Inset: the fluorescence image of one representative cell before and after the interaction with D-glucose. (**d**) Time-dependent fluorescence changes of HeLa cells that include functionalized QDs upon addition of 50 mM D-glucose to: (*a*) Nontreated HeLa cells, and (*b*) Taxol-treated HeLa cells. Each data point corresponds to the analysis of 20 different cells (Reproduced with permission from Freeman et al. 2009. Copyright Wiley-VCH Verlag GmbH & Co. KGaA)

is enhanced. Accordingly, different concentrations of tyrosinase could be analyzed by following the degree of quenching of the QDs, at a fixed time-interval of the biocatalytic oxidation of the functionalized QDs. Furthermore, the peptide (**3.1**) included the specific amino-acid sequence for hydrolytic cleavage by thrombin. Accordingly, the luminescence-quenched dopaquinone-functionalized QDs, generated by the oxidation of the modified QDs with tyrosinase, were interacted with thrombin. This cleaved the peptide and removed the dopaquinone residues from the QDs, a process that regenerated the luminescence properties of the QDs (Gill et al. 2006).

The application of the fluorescence resonance energy transfer (FRET) process to follow biocatalytic reactions is exemplified with the analysis of casein kinase, CK2, as a general platform for analyzing kinases (Freeman et al. 2010). These enzymes phosphorylate, in the presence of ATP, serine- or tyrosine-containing peptides to the respective phosphate esters. CdSe/ZnS semiconductor QDs were functionalized with the serine-containing peptide, (**3.2**), which is specific for CK2. Interaction of the chemically modified QDs with CK2 in the presence of Atto-590-modified ATP resulted in the formation of the dye-labeled phosphorylated product, Fig. 1.4C. The resulting FRET process, Fig. 1.4D, between the QDs and the dye was intensified as the phosphorylation was prolonged, demonstrating that the system enabled the probing of the dynamics of the phosphorylation reaction. A further configuration for the analysis of CK2 is shown in Fig. 1.4E that implements a FRET-based immunoassay approach. The antibody which is specific for the phosphorylated peptide was modified with the Atto-590 dye. The (**3.2**)-functionalized CdSe/ZnS QDs were subjected to CK2 and ATP in the presence of the dye-labeled antibody. The association of the antibody to the phosphorylated peptide led to a close proximity between the QDs and the FRET-acceptor dye, resulting in a FRET process in the complex structure. Figure 1.4F depicts the time-dependent luminescence changes in the system as a result of the phosphorylation process. At t=0, only the luminescence of the QDs is observed, and as the phosphorylation proceeds, the FRET signals are intensified.

The dynamic monitoring of intracellular metabolic processes and the application of the sensing platform for drug screening will be discussed here with the application of Nile blue, (**3.3**), (NB)-functionalized QDs acting as a sensor for the 1,4-dihydronicotin-amide cofactor. CdSe QDs were functionalized with Nile-blue (Freeman et al. 2009). The absorbance spectrum overlap of NB with the luminescence of the particles resulted in a FRET quenching of the QDs. The reduction of NB by the 1, 4-dihydronicotinamide, NADH resulted in the formation of the colorless reduced dye. This prohibited the FRET quenching of the QDs and restored their luminescence, Fig. 1.5A. The method enabled the use of the NB-modified QDs as a quantitative sensor for the detection of the NAD(P)H cofactor, Fig. 1.5B, and as a sensor for probing any $NAD(P)^+$-dependent enzyme and its substrate. Another important achievement included the use of the NB-functionalized QDs for monitoring the intracellular metabolism, and for probing the effect of drugs on the cell metabolism. The NB-modified QDs were incorporated into HeLa cancer cells via electroporation, and the cells were cultured under "starvation" conditions. The cells were, then, subjected to the addition of D-glucose that activated the cell metabolism. This resulted in the formation of the NAD(P)H cofactor that reduced the NB units, and triggered on the luminescence of the QDs. Figure 1.5C depicts the luminescence features of a specific cell before adding the D-glucose and after the activation of the cell metabolism. Clearly, the luminescence of the cell after treatment with glucose is enhanced. Figure 1.5C, curve (a), shows the time-dependent integrated light intensities generated by a collection of cells subjected to the addition of D-glucose. The constant increase in the luminescence intensity reflects the continuous reduction of the NB-units associated with the cells by the NAD(P)H cofactor generated by the cell metabolism. For comparison, Fig. 1.5C, curve (b), depicts the time-dependent fluorescence changes of the cells treated with L-glucose. Under these conditions no changes were observed, consistent with the fact that L-glucose is a non-native nutrient of the cells and does not activate their metabolism. The system was, then, implemented to examine the effect of taxol, (**3.4**), known as an anti-cancer drug that inhibits the metabolic pathways in cancer cells. Figure 1.5D, curve (a) shows the time-dependent luminescence changes of the QDs-labeled HeLa cells upon triggering the cell metabolism with D-glucose. For comparison, curve (b), depicts the effect of added D-glucose on the HeLa cells cultured under starvation conditions in the presence of taxol. Evidently, the luminescence changes are very low, suggesting that the addition of taxol inhibited the cell metabolism, as expected from its anti-drug activity (Freeman et al. 2009). These results are certainly a first step towards exciting future challenging applications of the functionalized QDs for screening drugs that effect cellular metabolism.

1.4 Biomolecules as Templates for Nanoscale Circuitry

The miniaturization of electronic circuits and the fabrication of dense electrical elements are continuous scientific challenges. In the past 40 years, the density of transistors per unit area was increased every 5 years by an order of magnitude by improving the lithographic methods that allowed the fabrication of the circuitries. It seems, however, that albeit the substantial recent progress in the 3D fabrication of nanometer-sized circuits, scientific limits start to be reached. We witness, however, in the past 10 years, scientific efforts to introduce a bottom-up approach to prepare nanocircuits. Although it is difficult to envisage at present the end-product of these activities, it is certain that creative and innovative research is emerging. The philosophy of the bottom-up approach defines atoms and molecules as the smallest objects that can be manipulated by chemical and physical means. Accordingly, the use of molecules or molecular aggregates as templates for the deposition of metallic or semiconductor circuitry, may provide means for the construction of nanometer-sized circuitry and devices. Realizing that this innovative concept might be a viable approach, the use of biomolecules as templates for the formation of the circuitry nanostructures

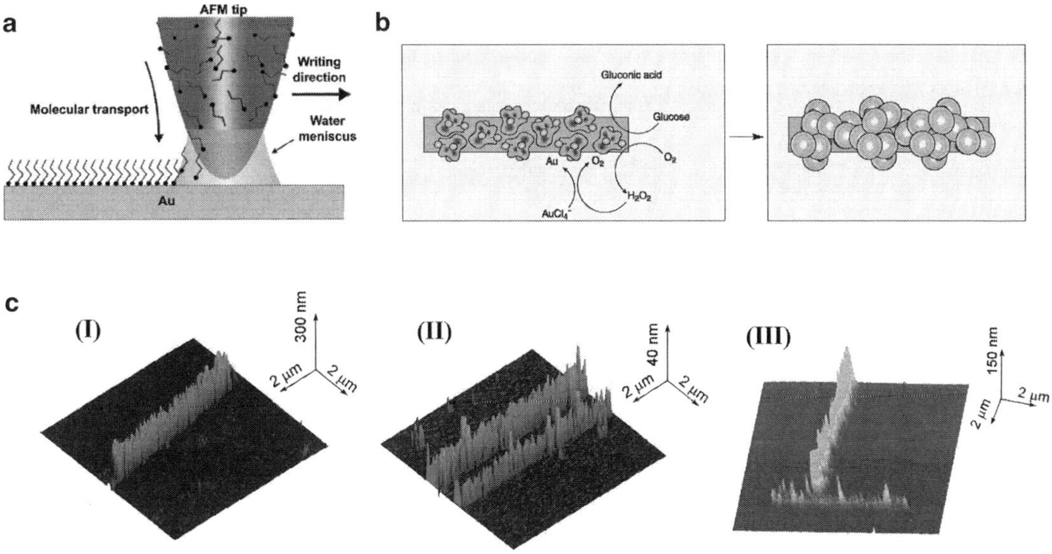

Fig. 1.6 (**a**) Basic working principle for DPN. (**b**) General scheme for the generation of a Au nanowire by the biocatalytic enlargement of a Au NP-functionalized GOx line deposited on a silicon support by DPN. (**c**) I – AFM image of the DPN pattern of GOx-Au nanowire after enlargement with glucose/$AuCl_4^-$. II – AFM image of Ag nanowires generated on two parallel AlkPh-Au templates deposited on the silicon support by DPN after enlargement in the Ag growth solution. III – AFM image of Au and Ag nanowires generated by deposition and Au enlargement of the Au NP-GOx, passivation with thiol, and subsequent deposition and Ag enlargement of the Au NP-AlkPh. *Part A* – Reproduced with permission from Ginger et al. (2004). Copyright Wiley-VCH Verlag GmbH & Co. KGaA. *Parts B and C* – Reproduced with permission from Basnar et al. (2006). Copyright Wiley-VCH Verlag GmbH & Co. KGaA

is an attractive choice. Biomolecules, proteins or nucleic acids self-assemble into one- or two-dimensional nanostructures, which exhibit encoded information that can provide the guided synthesis of metallic/semiconductor nanostructures. For example, actin filaments, DNA strands (λ-DNA or telomers), or amyloid microtubles provide nanometer-sized biopolymers on which the circuitry may be constructed. Furthermore, tremendous progress was achieved in designing DNA building blocks that self-assemble into 1D, 2D and 3D nanostructures (Simmel 2008; Niemeyer 2000; Lin et al. 2006). These systems may, then, provide nano-engineered playgrounds for the assembly of the electrical elements.

This topic will be introduced here by discussing several examples that implement proteins as templates for the preparation of nanowires and as catalysts for growing metallic nanowires. Dip-pen-nanolithography (DPN) provides a useful method to deposit nanometer-sized chemical patterns on surfaces (Ginger et al. 2004; Wu et al. 2011; Basnar and Willner 2009; Salaita et al. 2007). The method applies the AFM tip as a nanoscale "fountain-pen" to write on surface, Fig. 1.6A. The tip is coated with the chemical ink, and upon the scanning of the tip along the surface, the ink is delivered to the support by capillary forces, thus generating a pattern. The DPN deposition of a catalytically-active pattern which stimulates the synthesis of metals or semiconductors, can then be used to prepare the nanocircuits. For example, the biocatalytic growth of metallic nanoparticles was used to synthesize metallic nanowires on surfaces (Basnar et al. 2006). Glucose oxidase (GOx) or alkaline phosphatase (AlkPh) were modified with Au NPs (1.4 nm diameter), and the enzyme/NPs hybrids were used to enlarge metallic nanowires. The Au NP/GOx hybrid was deposited on a silicon surface (Basnar et al. 2006) using the DPN patterning method. The resulting biocatalytically-patterned surface was subjected to the developing solution that consisted of glucose/$AuCl_4^-$. The enzyme-catalyzed oxidation of glucose yielded H_2O_2, which reduced the $AuCl_4^-$ on the Au NPs seeds, Fig. 1.6B. The enlargement of the NPs led to the formation of contacted Au NPs aggregates that formed a long conductive Au

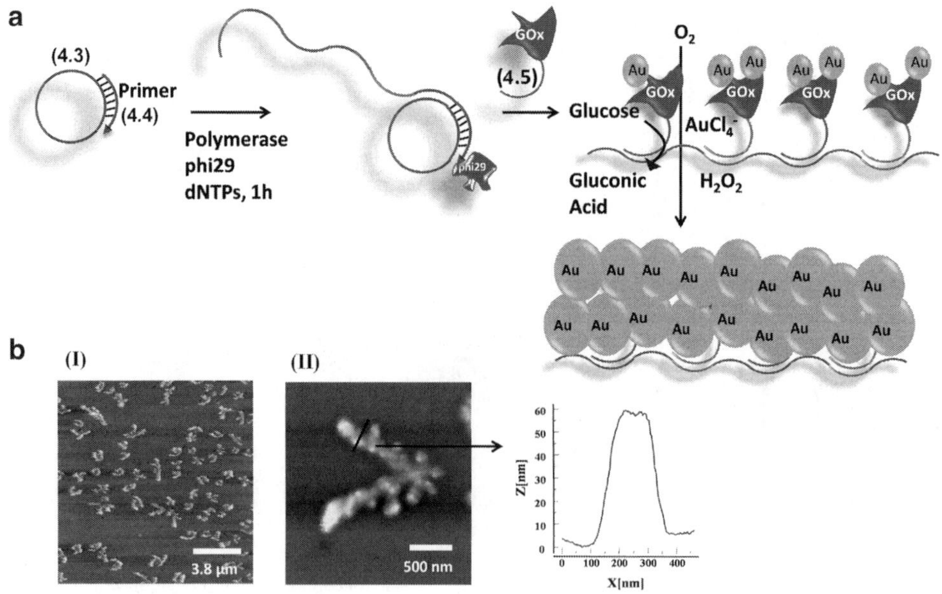

Fig. 1.7 (**a**) Assembly of Au NP-functionalized GOx on the RCA-synthesized template, and the biocatalytic enlargement of the NPs to Au nanowire structures. (**b**) AFM images of the resulting Au nanostructures: (I) A large 12 μm^2 image. (II) A larger magnification image of a Au nanowire and the respective cross-section analysis. *Part B* – Reprinted with permission from Wilner et al. (2009). Copyright (2009) American Chemical Society

nanowire with a width of ca. 80 nm, Fig. 1.6C, image (I). Similarly, the Au NPs/AlkPh hybrid was deposited on the silicon surface. This enzyme catalyzed the hydrolysis of p-amino phenol phosphate, (**4.1**), to p-amino phenol, (**4.2**). The (**4.2**) product acts as a reducing agent for the reduction of Ag^+ ions on the Au NPs catalytic seeds, leading to core-shell Au/Ag enlarged NPs, and their aggregation to electrically contacted Ag° nanowires. Figure 1.6C, Image (II), shows the formation of 25–30 nm wide silver wires. Furthermore, by the DPN-deposition of the Au NP/GOx and the Au NP/AlkPh patterns, with the sequential treatment of the surface with the respective developing solutions (glucose/ $AuCl_4^-$ or (**4.1**)/Ag^+), the orthogonal deposition of two different metal nanowires (Au and Ag) was demonstrated, Fig. 1.6C, Image (III). One could, however, argue that chemically enlarged DPN-patterned metal nanoparticles could similarly yield metal nanowires, and, thus, the need for the biocatalytic growth of the nanowire might be raised. Besides the possibility to orthogonally synthesize nanowires of different compositions, the biocatalytic growth of the nanowires leads to controlled, high-quality, nanowire structures. While the chemical growth of nanoparticles is a random process that can only be stopped by the physical removal of the developing solution, the biocatalytic growth of the nanowires is a controlled process which is accompanied by a self-inhibition mechanism. That is, the reducing agent is generated on the biocatalytic pattern, and as the catalytic particles are enlarged, the diffusion of the substrate to the enzyme is perturbed. Thus, the coating of the enzyme patterns by the metallic nanoparticles inhibits the enzymatic functions and the growth of the nanowires is blocked. Thus, the random growth of the nanowires is eliminated. Furthermore, the dimensions of the resulting nanowires are controlled by the dimensions of the biocatalysts, and for example, the width of the nanowire generated by GOx corresponded to 80–100 nm, whereas the width of the Ag nanowire synthesized by the smaller AlkPh biocatalyst corresponded to 25–30 nm.

A further example demonstrating the biocatalytic growth of metallic nanowires has involved the use of DNA chains as templates for the assembly of the enzyme units (Wilner et al. 2009), Fig. 1.7A. Using a circular DNA, (**4.3**), and a primer nucleic acid, (**4.4**), the polymerase stimulated rolling circle amplification (RCA) process that was activated in the presence of the nucleotide mixture, dNTPs, to yield micrometer-

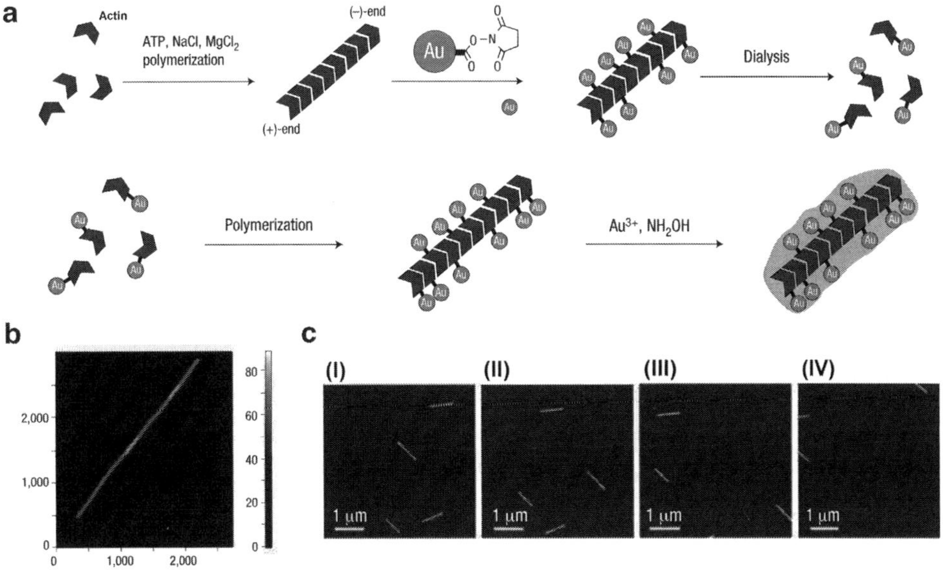

Fig. 1.8 (**a**) Scheme for the synthesis of the actin-based metallic nanowires. The shading on the final image indicates continuous wire formation. (**b**) AFM image of the Au nanowire generated on the actin template. All dimensions are in nanometers. (**c**) ATP-fuelled motility of the actin-Au nanoblock-actin filaments on a myosin interface. Four consecutive microscopy images (I–IV), taken at intervals of 5 s, are shown, corresponding to the motility of the actin-Au nanoblock-actin nanotransporter associated with the myosin interface after the addition of ATP (Reproduced with permission from Patolsky et al. 2004. Copyright Nature Publishing Group, 2004)

long DNA chains consisting of repeat units complementary to the RCA template. Glucose oxidase modified with Au NPs (1.4 nm diameter) was further functionalized with a nucleic acid (**4.5**) that was complementary to a domain of the repeat units of the RCA product. The resulting enzyme/Au NPs hybrid was hybridized on the RCA chain, and in the presence of glucose and $AuCl_4^-$ the biocatalytic growth of the Au nanowires was activated. Accordingly, the biocatalyzed oxidation of glucose generated gluconic acid and H_2O_2, and the resulting H_2O_2 reduced $AuCl_4^-$ on the Au NPs seeds to form the Au nanowires. The resulting micrometer-long Au nanowires exhibited a height corresponding to ca. 60 nm, Fig. 1.7B. Other approaches have implemented proteins as nanoreactors for the synthesis of nanowires. For example, oligodiphenyalanine nanotubules were used as nanoreactors for the chemical synthesis of Ag nanowires that were purified from the protein coating layer by the enzymatic digestion of the nanotubules.

The protein-templated synthesis of metallic nanowires enabled also the fabrication of nanoscale motor devices (Patolsky et al. 2004a). For example, g-actin monomer units were polymerized with ATP in the presence of Mg^{2+} ions, and the resulting actin filaments were functionalized with Au NPs (1.4 nm), Fig. 1.8A. The resulting filaments were, then, dissociated by dialyzing the ATP/Mg^{2+} to yield g-actin monomers modified with the single Au NPs. The subsequent polymerization of the g-actin/Au NP hybrid units in the presence of ATP/Mg^{2+} yielded the Au NPs filaments that were further subjected to react with base g-actin units. This led to hybrid patterned actin filaments, where the central Au NP-functionalized domain was conjugated at its ends to actin filaments. The subsequent chemical enlargement of the Au nanoparticles yielded micrometer long Au nanowires (width 80–100 nm) that were linked at their ends to f-actin filaments, Fig. 1.8B. The actin-myosin couple is a central motor protein. It is responsible for numerous mechanical functions in biological systems such as muscle operation or intracellular transport of biomolecular components. The myosin is a Y-shaped enzyme that binds through its anchoring sites to the actin filament. The ATP fuel binds to the groove generated in-between the arms of the linked myosin. Hydrolysis of ATP to ADP is associated with the dissociation of one of the myosin arms and its translocation on the actin filament. By the repeated hydrolysis of the

ATP fuel, the directional motility of myosin proceeds. The successful synthesis of the Au nanowire conjugated at its ends to f-actin was used to develop a nanoscale ATP-fueled motor. A glass slide was modified with a monolayer of myosin, and the Au-wire/f-actin nano-objects were deposited on the surface to yield surface confined actin-filament/myosin complexes. Upon the addition of ATP, the motility of the Au nanowire was demonstrated using reflectance microscopy. Figure 1.8C shows the moving elements at time interval of 5 s on the frame of the same area. The nanowires move at a speed of 2.5 $nm{\cdot}s^{-1}$. The nano-objects move as long as the ATP fuel is present in the system, and upon its consumption, regenerate their rest position. Re-addition of ATP restores the motility of the objects on the surface.

1.5 Conclusions and Perspectives

The present article has introduced several evidences for the rapidly developing area of nanobiotechnology. The conjugation between nanomaterials (metallic nanoparticles, carbon nanotubes or semiconductor quantum dots) with biomolecules yields new hybrid materials of new functions and properties. The unique optical, electronic and catalytic properties of nanomaterials can be integrated with the recognition and catalytic properties of biomolecules and provide new methods to develop new kinds of electronic and optical biosensors. Fundamental issues, such as the electrical contacting of redox proteins with electrodes, the multiplexed analysis of analytes or the probing of the dynamics of biotransformations were demonstrated with these systems. Particularly interesting are recent efforts to use these hybrid nanostructures for the imaging of intracellular metabolic processes. While substantial progress was demonstrated in the area of nanobiotechnology, important challenges are ahead of us. The rapid progress in the discovery of new plasmonic phenomena, and particularly nanophotonics suggests that new plasmonic-based sensing platforms will emerge in the near future. Furthermore, the research addressing the incorporation of biomolecule/nanoparticle hybrids into cellular environments is still at its infancy. The unique optical and catalytic properties of nanoparticles are, however, anticipated to provide new means to follow intracellular metabolic processes and transport phenomena. Furthermore, nanoparticles may act as carriers or support for delivering drugs or chemical components that silence or activate intracellular processes. All these possibilities will not only provide intracellular sensing, but are also expected to introduce new nano-medical applications including therapeutic treatment or drug screening.

Finally, the use of biomolecules as templates for the bottom-up assembly of nanoscale objects and devices holds a great promise for the miniaturization of functional systems. Specifically, the conjugation of enzymes or motor proteins to nanoscale wires or tubes could lead to new nanoscale delivery units and mechanical sensors. Realizing the impressive progress in the area of nanobiotechnology in the past decade, we feel confident that this interdisciplinary research area will continue to attract substantial scientific interest.

Acknowledgements The support of the Israel Science Foundation is acknowledged.

References

Alivisatos, A. P. (1996). Semiconductor clusters, nanocrystals, and quantum dots. *Science, 271*, 933–937.

Alivisatos, A. P. (2003). Room-temperature single-nucleotide polymorphism and multiallele DNA detection using fluorescent nanocrystals and microarrays. *Analytical Chemistry, 75*, 4766–4772.

Baron, R., Zayats, M., & Willner, I. (2005). Dopamine-, L-DOPA-, adrenaline- and noradrenaline-induced growth of au-nanoparticles: Assays for the detection of neurotransmitters and of tyrosinase activity. *Analytical Chemistry, 77*, 1566–1571.

Bartlett, P. N., Tebbutt, P., & Whitaker, R. C. (1991). Kinetic aspects of the use of modified electrodes and mediators in bioelectrochemistry. *Progress in Reaction Kinetics, 16*, 55–155.

Barton, S. C., Gallaway, J., & Atanassov, P. (2004). Enzymatic biofuel cells for implantable and microscale devices. *Chemical Reviews, 104*, 4867–4886.

Basnar, B., & Willner, I. (2009). Dip-pen-nanolithographic patterning of metallic, semiconductor, and metal oxide nanostructures on surfaces. *Small, 5*, 28–44.

Basnar, B., Weizmann, Y., Cheglakov, Z., & Willner, I. (2006). Synthesis of nanowires using dip-pen nanolithography and biocatalytic inks. *Advanced Materials, 18*, 713–718.

Freeman, R., Gill, R., Shweky, I., Kotler, M., Banin, U., & Willner, I. (2009). Biosensing and probing of intracellular metabolic pathways by NADH-sensitive quantum dots. *Angewandte Chemie (International ed. in English), 48*, 309–313.

Freeman, R., Finder, T., Gill, R., & Willner, I. (2010). Probing protein kinase (CK2) and alkaline phosphatase with CdSe/ZnS quantum dots. *Nano Letters, 10*, 2192–2196.

Gerion, D., Chen, F. Q., Kannan, B., Fu, A. H., Parak, W. J., Chen, D. J., Majumdar, A., Willard, D. M., Carillo, L. L., Jung, J., & Van Orden, A. (2001). CdSe-ZnS quantum dots as resonance energy transfer donors in a model protein-protein binding assay. *Nano Letters, 1*, 469–474.

Gill, R., Freeman, R., Xu, J. P., Willner, I., Winograd, S., Shweky, I., & Banin, U. (2006). Probing biocatalytic transformations with CdSe-ZnS QDs. *Journal of the American Chemical Society, 128*, 15376–15377.

Gill, R., Zayats, M., & Willner, I. (2008). Semiconductor quantum dots for bioanalysis. *Angewandte Chemie (International ed. in English), 47*, 7602–7625.

Ginger, D. S., Zhang, H., & Mirkin, C. A. (2004). The evolution of dip-pen nanolithography. *Angewandte Chemie (International ed. in English), 43*, 30–45.

Goldman, E. R., Clapp, A. R., Anderson, G. P., Uyeda, H. T., Mauro, J. M., Medintz, I. L., & Mattoussi, H. (2004). Multiplexed toxin analysis using four colors of quantum dot fluororeagents. *Analytical Chemistry, 76*, 684–688.

Heller, A. (1992). Electrical connection of enzyme redox centers to electrodes. *The Journal of Physical Chemistry, 96*, 3579–3587.

Heller, A. (2004). Miniature biofuel cells. *Physical Chemistry Chemical Physics: PCCP, 6*, 209–216.

Katz, E., & Willner, I. (2004). Integrated nanoparticle–biomolecule hybrid systems: Synthesis, properties, and applications. *Angewandte Chemie (International ed. in English), 43*, 6042–6108.

Kelly, K. L., Coronado, E., Zhao, L. L., & Schatz, G. C. (2003). The optical properties of metal nanoparticles: The influence of size, shape, and dielectric environment. *The Journal of Physical Chemistry. B, 107*, 668–67.

Lin, C., Liu, Y., Rinker, S., & Yan, H. (2006). DNA tile based self-assembly: Building complex nanoarchitectures. *ChemPhysChem, 7*, 1641–1647.

Maidan, R., & Heller, A. (1992). Elimination of electrooxidizable interferant-produced currents in amperometric biosensors. *Analytical Chemistry, 64*, 2889–2896.

Marcus, R. A., & Sutin, N. (1985). Electron transfers in chemistry and biology. *Biochimica et Biophysica Acta, 811*, 265–322.

Mattoussi, H., Mauro, J. M., Goldman, E. R., Anderson, G. P., Sundar, V. C., Mikulec, F. V., & Bawendi, M. G. (2000). Self-assembly of CdSe-ZnS quantum dot bioconjugates using an engineered recombinant protein. *Journal of the American Chemical Society, 122*, 12142–12150.

Medintz, I. L., Uyeda, H. T., Goldman, E. R., & Mattoussi, H. (2005). Quantum dot bioconjugates for imaging, labelling and sensing. *Nature Materials, 4*, 435–446.

Medintz, I. L., Clapp, A. R., Brunel, F. M., Tiefenbrunn, T., Uyeda, H. T., Chang, E. L., Deschamps, J. R., Dawson, P. E., & Mattoussi, H. (2006). Proteolytic activity monitored by fluorescence resonance energy transfer through quantum-dot–peptide conjugates. *Nature Materials, 5*, 581–589.

Murphy, L. (2006). Biosensors and bioelectrochemistry. *Current Opinion in Chemical Biology, 10*, 177–184.

Niemeyer, C. M. (2000). Self-assembled nanostructures based on DNA: Towards the development of nanobiotechnology. *Current Opinion in Chemical Biology, 4*, 609–661.

Niemeyer, C. M. (2001). Nanoparticles, proteins, and nucleic acids: Biotechnology meets materials science. *Angewandte Chemie (International ed. in English), 40*, 4128–4158.

Nirmal, M., & Brus, L. (1999). Luminescence photophysics in semiconductor nanocrystals. *Accounts of Chemical Research, 32*, 407–414.

Park, S.-J., Taton, T. A., & Mirkin, C. A. (2002). Array-based electrical detection of DNA with nanoparticle probes. *Science, 295*, 1503–1506.

Patolsky, F., Weizmann, Y., & Willner, I. (2004a). Designing actin-based metallic nanowires and bio-nanotransporters. *Nature Materials, 3*, 692–695.

Patolsky, F., Weizmann, Y., & Willner, I. (2004b). Long-range electrical contacting of redox-enzymes by single-walled carbon nanotube connectors. *Angewandte Chemie (International ed. in English), 43*, 2113–2117.

Rajagopalan, R., Aoki, A., & Heller, A. (1996). Effect of quaternization of the glucose oxidase "wiring" redox polymer on the maximum current densities of glucose electrodes. *The Journal of Physical Chemistry, 100*, 3719–3729.

Salaita, K., Wang, Y., Vega, R. A., & Mirkin, C. A. (2007). Applications of dip-pen nanolithography. *Nature Nanotechnology, 2*, 145–155.

Schuhmann, W., Ohara, T. J., Schmidt, H.-L., & Heller, A. (1991). Electron transfer between glucose oxidase and electrodes via redox mediators bound with flexible chains to the enzyme surface. *Journal of the American Chemical Society, 113*, 1394–1397.
Shipway, A. N., Katz, E., & Willner, I. (2000). Nanoparticle arrays on surfaces for electronic, optical and sensoric applications. *ChemPhysChem, 1*, 1–208.
Simmel, F. C. (2008). Three-dimensional nanoconstruction with DNA. *Angewandte Chemie (International ed. in English), 47*, 5884–5887.
Wang, J. (2008). Electrochemical glucose biosensors. *Chemical Reviews, 108*, 814–882.
Willner, I., & Willner, B. (2001). Biomaterials integrated with electronic elements: En route to bioelectronics. *Trends in Biotechnology, 19*, 222–230.
Willner, I., Lapidot, N., Riklin, A., Kasher, R., Zahavy, E., & Katz, E. (1994). Electron transfer communication in glutathione reductase assemblies: Electrocatalytic, photocatalytic and catalytic systems for the reduction of oxidized glutathione. *Journal of the American Chemical Society, 116*, 1428–1441.
Willner, I., Heleg-Shabtai, V., Blonder, R., Katz, E., Tao, G., Bückmann, A. F., & Heller, A. (1996). Electrical wiring of glucose oxidase by reconstitution of FAD-modified monolayers assembled onto Au-electrodes. *Journal of the American Chemical Society, 118*, 10321–10322.
Willner, B., Katz, E., & Willner, I. (2006a). Electrical contacting of redox proteins by nanotechnological means. *Current Opinion in Biotechnology, 17*, 589–596.
Willner, I., Baron, R., & Willner, B. (2006b). Growing metal nanoparticles by enzymes. *Advanced Materials, 18*, 1109–1120.
Willner, I., Yan, Y.-M., Willner, B., & Tel-Vered, R. (2009). Integrated enzyme-based biofuel cells–A review. *Fuel Cells, 9*, 7–24.
Wilner, O. I., Shimron, S., Weizmann, Y., Wang, Z.-G., & Willner, I. (2009). Self-assembly of enzymes on DNA scaffolds: En route to biocatalytic cascades and the synthesis of metallic nanowires. *Nano Letters, 9*, 2040–2043.
Wu, C.-C., Reinhoudt, D. N., Otto, C., Subramaniam, V., & Velders, A. H. (2011). Patterning: Strategies for patterning biomolecules with dip-pen nanolithography. *Small, 7*, 989–1002.
Xiao, Y., Patolsky, F., Katz, E., Hainfeld, J. F., & Willner, I. (2003). "Plugging into enzymes": Nanowiring of redox enzymes by a gold nanoparticle. *Science, 299*, 1877–1881.
Xiao, Y., Pavlov, V., Levine, S., Niazov, T., Markovitch, G., & Willner, I. (2004). Catalytic growth of au-nanoparticles by NAD(P)H cofactors: Optical sensors for NAD(P)+–dependent biocatalyzed transformations. *Angewandte Chemie (International ed. in English), 43*, 4519–4522.
Zayats, M., Katz, E., & Willner, I. (2002). Electrical contacting of flavoenzymes and NAD(P)+–dependent enzymes by reconstitution and affinity interactions on phenylboronic acid monolayers associated with au-electrodes. *Journal of the American Chemical Society, 124*, 14724–14735.
Zayats, M., Baron, R., Popov, I., & Willner, I. (2005). Biocatalytic growth of Au nanoparticles: From mechanistic aspects to biosensors design. *Nano Letters, 5*, 21–25.

Chapter 2
Superresolution Optical Fluctuation Imaging (SOFI)

Thomas Dertinger, Ryan Colyer, Robert Vogel, Mike Heilemann, Markus Sauer, Jörg Enderlein, and Shimon Weiss

Abstract Superresolution microscopy has shifted the limits for fluorescence microscopy in cell biology. The possibility to image cellular structures and dynamics of fixed and even live cells and organisms at resolutions of several nanometers holds great promise for future biological discoveries. We recently introduced a novel superresolution technique, based on the statistical evaluation of stochastic fluctuations stemming from single emitters, dubbed "superresolution optical fluctuation imaging" (SOFI). In comparison to previously introduced superresolution methods, SOFI exhibits favorable attributes such as simplicity, affordability, high speed, and low levels of light exposure. Here we summarize the basic working principle and recent advances.

Keywords Superresolution • Statistical analysis • Correlation function

2.1 Main Article

Diffraction-based imaging has limited resolving power. In 1873, Ernst Abbe related the determining parameters, such as the wavelength of light, to the resolving power of microscopes, i.e. to what extent the signal stemming from a point source will be blurred due to diffraction of the optical

T. Dertinger (✉) • R. Colyer • R. Vogel
Department of Chemistry and Biochemistry, University of California Los Angeles, Los Angeles, CA, USA
e-mail: thomasd@chem.ucla.edu

M. Heilemann
Department of Physics, Applied Laser Physics, Bielefeld University, Bielefeld, Germany

M. Sauer
Department of Biotechnology and Biophysics, Julius-Maximilians-Universität Würzburg, Würzburg, Germany

J. Enderlein
III Institute of Physics, Georg August University, Göttingen, Germany

S. Weiss
Department of Chemistry and Biochemistry, University of California Los Angeles, Los Angeles, CA, USA

Department of Physiology, University of California Los Angeles, UCLA, Los Angeles, CA, USA

California NanoSystems Institute, University of California Los Angeles, UCLA, Los Angeles, CA, USA

E. Zahavy et al. (eds.), *Nano-Biotechnology for Biomedical and Diagnostic Research*,
Advances in Experimental Medicine and Biology 733, DOI 10.1007/978-94-007-2555-3_2,

system (Abbe 1873). For visible light, the maximum achievable resolution in far-field imaging is around 250 nm. Smaller structures will not be resolved but appear as a blurred spot, the Point Spread Function (PSF). Fluorescence far-field imaging is one of the most important tools for studying live cells, tissues and small organisms. The high contrast and sensitivity afforded by fluorescence imaging, the site-specific targetability of relatively small exogenous fluorescence probes, the facile genetic manipulation and specific labelling by fluorescence proteins fusions, and the ability to multiplex the detection by multi-colour imaging allows for relating structure, function and dynamics in live organisms. Over many decades it was commonly believed that complementary (to fluorescence) methods must be used if higher resolution was required. These methods, such as electron microscopy (EM), atomic force microscopy (AFM), or near-field scanning optical microscopy (NSOM) in turn lack the penetration depth, involve often elaborate sample preparation, and in most cases are performed on a fixed specimen.

The ability to overcome the resolution limit set by diffraction is called superresolution imaging. Various powerful methods, such as stimulated emission depletion (STED) (Hell and Wichmann 1994), (fast) photoactivated localization microscopy (f)PALM (Betzig et al. 2006; Hess et al. 2006), stochastic optical reconstruction microscopy STORM (Rust et al. 2006), directSTORM (Heilemann et al. 2008) and ground state depletion imaging GSDIM (Fölling et al. 2008), have been developed during the last decade. Fundamentally, all superresolution imaging techniques are relying on a transition from a fluorescent ('on') to a non-fluorescent ('off') state of the probe, or on the use of entangled photons (Mitchell et al. 2004; Walther et al. 2004). With superresolution microscopy it is now possible to take advantage of all features of conventional fluorescence microscopy with a dramatically increased resolution. For the first time, synaptic vesicles inside the axons of cultured neurons could be imaged at video-rate at a resolution of 60 nm using STED (Westphal et al. 2008). Recently, histone H2B core proteins in the nucleus of live HeLa cells have been monitored featuring a resolution of ~20 nm (Wombacher et al. 2010). Already in 2008, superresolution imaging in *Caulobacter crescentus* at 40 nm has been demonstrated using PALM (Biteen et al. 2008). The achieved resolution produced images of unprecedented clarity and pushed the frontier of fluorescence imaging to previously inaccessible terrain.

Superresolution comes at a cost. Besides the monetary aspect, the speed and high resolution of STED for example has to be paid with a high light exposure of the sample, leading to the build-up of phototoxic products and eventually cell death. On the other hand, high resolution in combination with low illumination intensities come at the expense of long acquisition times as seen in stochastic photoswitching methods such as (f) PALM, (d) STORM, rendering these methods sometimes incapable of live cell imaging.

Recently, we developed a novel superresolution technique, superresolution optical fluctuation imaging (SOFI), which features a moderate resolution gain within short acquisition times and medium light exposure levels – a trade-off that other superresolution techniques cannot provide (Dertinger et al. 2009). SOFI trades-off ultrahigh superresolution for shorter acquisition time and lower light level exposure. Other favorable attributes of SOFI include resolution enhancement along all three dimensions (in the plane and along the optical axis) and elimination of uncorrelated background signal (i.e. contrast enhancement). Furthermore, SOFI can be performed on all imaging platforms without modifications (Fig. 2.1).

SOFI is a purely software-based technique. Thus, the choice of the imaging platform and signal can be carefully optimized for a given research task. For SOFI to work, sub-diffraction sized emitters need to fluctuate independently. These fluctuations could stem from different molecular origins such as reorientations (detected by polarization optics), inter-system crossing to the triplet-states, and even fluctuations of non-fluorescent signals such as light scattering from re-orienting gold nanorods (as detected by polarized dark field microscope). The only prerequisite is that one has to be able to image as fast as the characteristic fluctuation time (even this obstacle could be theoretically overcome, by clever illumination schemes).

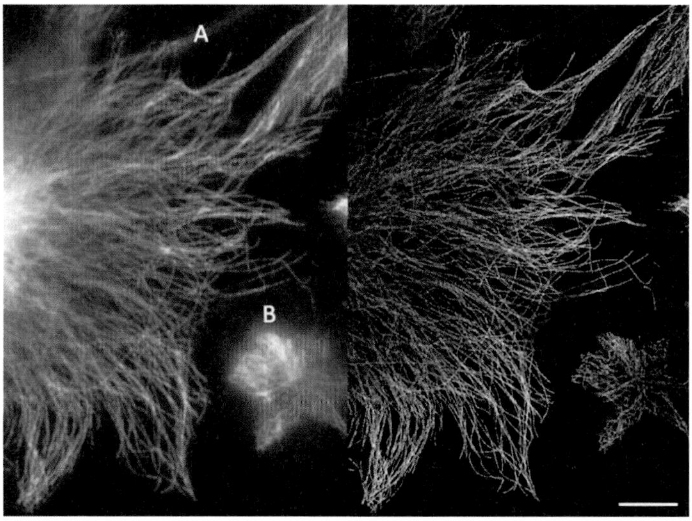

Fig. 2.1 Tubulin network of a 3T3 fibroblast labeled with QD800 infrared emitting quantum dots, imaged on a conventional widefield setup using a Xenon lamp as illumination source. *Left*: original widefield image. *Right*: Second-order SOFI image. The resolution enhancement is readily apparent. The intrinsic resolution enhancement along 3D of SOFI results in optical sectioning of a 2D image. Hence the out-of-focus feature indicated by the letter 'A' disappears in the SOFI image. The background elimination can be appreciated by comparing the region around the letter 'B' to the same region in the SOFI image. Scalebar: 10 μm. Acquisition time 100 s (0.1 s/frame)

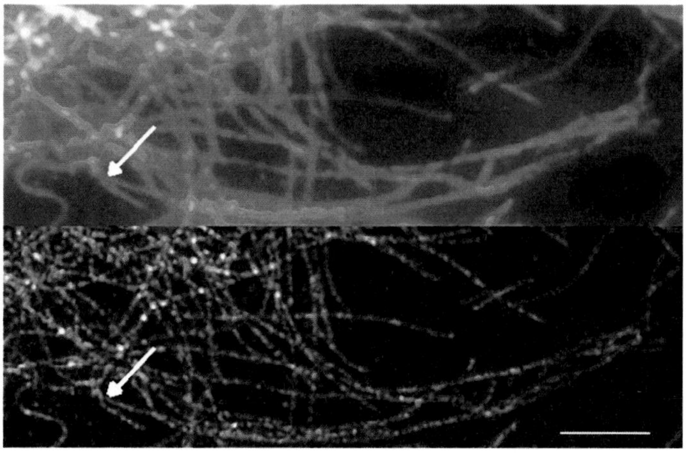

Fig. 2.2 Tubulin network of COS-7 cells labeled with Alexa 647 conjugated antibodies. *Upper panel:* Original fluorescence image as obtained by averaging over 200 frames. *Lower panel:* SOFI image as obtained by the analysis of the same 200 frames (~4.5 s). The *arrows* point to a feature which can be resolved only in the SOFI image, demonstrating the resolution gained by SOFI scalebar 5 μm

SOFI harnesses the power of higher-order cumulants to produce superresolution images. Provided sufficient photon statistics have been acquired, the higher the order of the cumulant calculated the higher the resulting resolution. While quantum dots are regarded as problematic in other superresolution techniques, for SOFI quantum dots are almost perfectly suited. Recently, we could show that not only quantum dots can be used as probes for SOFI imaging, but also conventional organic dyes (Dertinger et al. 2010b). This achievement marks an important step forward in terms of a broad applicability of SOFI for bioimaging (Fig. 2.2).

The SOFI algorithm allows for the generation of virtual pixels by cross-correlating adjacent pixels of the camera (Dertinger et al. 2010a). This in turn dramatically improves the performance of SOFI, since the effective pixel size of the imaging system shrinks to the same degree as the resolution is enhanced.

For example, an initial pixel size of 160 nm for the original imaging system, will yield a SOFI image consisting of four-times more pixels, each of them having an effective size of 80 nm while featuring a PSF which is twice as small. The PSF oversampling ratio (width of the PSF/size of a pixel) of the imaging system therefore remains constant even for the superresolved image. This also holds true for higher-order cumulant images.

2.2 Conclusion

SOFI offers the ability to perform superresolution imaging with great flexibility with regard to the imaging setup and the labeling probe. It has the potential to reach sub-second acquisition times. The inherent background reduction and optical sectioning capabilities allow for optical sectioning even on a wide field microscope. SOFI can be performed on any microscope, and depending on the data quality, images of increasing resolution can be calculated. This in turn offers great flexibility in acquisition, since the resolution enhancement has not to be set in advance, but can be adjusted post acquisition.

Acknowledgments This work was supported by NIH grant# 5R01EB000312 and NIH grant# 1R01GM086197. Thomas Dertinger is supported by the German Science Foundation (DFG, fellowship # DE 1591/1 1). Jörg Enderlein acknowledges financial support by the Human Frontier Science Program (RGP46/2006) and by the German Federal Ministry of Education and Research (FKZ 13N9236).

References

Abbe, E. (1873). Contributions to the theory of the microscope and the microscopic perception (Translated from German). *Archiv Fur Mikroskopische Anatomic, 9*, 413–468.

Betzig, E., Patterson, G. H., Sougrat, R., Lindwasser, O. W., Olenych, S., Bonifacino, J. S., et al. (2006). Imaging intracellular fluorescent proteins at nanometer resolution. *Science, 313*(5793), 1642–1645. doi:10.1126/science.1127344.

Biteen, J. S., Thompson, M. A., Tselentis, N. K., Bowman, G. R., Shapiro, L., & Moerner, W. E. (2008). Superresolution imaging in live Caulobacter crescentus cells using photoswitchable EYFP. *Nature Methods, 5*(11), 947–949. doi:10.1038/nmeth.1258.

Dertinger, T., Colyer, R., Iyer, G., Weiss, S., & Enderlein, J. (2009). Fast, background-free, 3D super-resolution optical fluctuation imaging (SOFI). *Proceedings of the National Academy of Sciences of the United States of America, 106*(52), 22287–22292. doi:10.1073/pnas.0907866106.

Dertinger, T., Colyer, R., Vogel, R., Enderlein, J., & Weiss, S. (2010a). Achieving increased resolution and more pixels with Superresolution Optical Fluctuation Imaging (SOFI). *Optics Express, 18*(18), 18875. doi:10.1364/OE.18.018875.

Dertinger, T., Heilemann, M., Vogel, R., Sauer, M., & Weiss, S. (2010b). Superresolution optical fluctuation imaging with organic dyes. *Angewandte Chemie (International ed. in English), 49*(49), 9441–9443. doi:10.1002/anie.201004138.

Fölling, J., Bossi, M., Bock, H., Medda, R., Wurm, C. A., Hein, B., et al. (2008). Fluorescence nanoscopy by ground-state depletion and single-molecule return. *Nature Methods, 5*(11), 943–945. doi:10.1038/nmeth.1257.

Heilemann, M., van de Linde, S., Schuttpelz, M., Kasper, R., Seefeldt, B., Mukherjee, A., et al. (2008). Subdiffraction-resolution fluorescence imaging with conventional fluorescent probes. *Angewandte Chemie (International ed. in English), 47*(33), 6172–6176. doi:10.1002/anie.200802376.

Hell, S., & Wichmann, J. (1994). Breaking the diffraction resolution limit by stimulated emission: Stimulated-emission-depletion fluorescence microscopy. *Optics Letters, 19*(11), 780–782. http://ol.osa.org/abstract.cfm?URI=ol-19-11-780.

Hess, S. T., Girirajan, T. P. K., & Mason, M. D. (2006). Ultra-high resolution imaging by fluorescence photoactivation localization microscopy. *Biophysical Journal, 91*(11), 4258–4272. doi:10.1529/biophysj.106.091116.

Mitchell, M. W., Lundeen, J. S., & Steinberg, A. M. (2004). Super-resolving phase measurements with a multiphoton entangled state. *Nature, 429*(6988), 161–164. doi:10.1038/nature02493.

Rust, M. J., Bates, M., & Zhuang, X. (2006). Sub-diffraction-limit imaging by stochastic optical reconstruction microscopy (STORM). *Nature Methods, 3*(10), 793–795. doi:10.1038/nmeth929.

Walther, P., Pan, J.-W., Aspelmeyer, M., Ursin, R., Gasparoni, S., & Zeilinger, A. (2004). De Broglie wavelength of a non-local four-photon state. *Nature, 429*(6988), 158–161. doi:10.1038/nature02552.

Westphal, V., Rizzoli, S. O., Lauterbach, M. A., Kamin, D., Jahn, R., & Hell, S. (2008). Video-rate far-field optical nanoscopy dissects synaptic vesicle movement. *Science, 320*(5873), 246–249. doi:10.1126/science.1154228.

Wombacher, R., Heidbreder, M., van de Linde, S., Sheetz, M. P., Heilemann, M., Cornish, V. W., et al. (2010). Live-cell super-resolution imaging with trimethoprim conjugates. *Nature Methods*. doi:10.1038/nmeth.1489.

Chapter 3
Application of Nanoparticles for the Detection and Sorting of Pathogenic Bacteria by Flow-Cytometry

Eran Zahavy, Raphael Ber, David Gur, Hagar Abramovich, Esti Freeman, Sharon Maoz, and Shmuel Yitzhaki

Abstract In this paper we will describe a new developed contribution of fluorescence nano-crystal (q-dots) as a fluorescence label for detecting pathogenic bacteria by flow cytometry (FCM) and the use of nano-magnetic particles to improve bacterial sorting by Flow cytometry cell sorting (FACS).

FCM or FACS systems are based upon single cell detection by light scatter and Immunofluorescence labeling signals. The common FACS systems are based upon single or dual excitation as excitation source both for light scatter parameters and for several fluorescence detectors. Hence, for multi-labeling detection, there is a need for fluorophores with broad excitation wave length and sharp emission bands. Moreover, such fluorophores should be with high fluorescence efficiency, stable, and available for bio-molecules conjugation. Q-dots benefit from practical features which meet those criteria. We will describe the use of q-dots as fluorescence labels for specific conjugates against *Bacillus anthracis* spores and *Yersinia pestis* bacteria, which enable the specific detection of the different species. A specific and sensitive multiplex analysis procedure for both pathogens was achieved, with high sensitivity down to 10^3 bacteria per ml in the sample.

Sorting bacteria by FACS has a tremendous advantage for sensitive and selective analysis and sorting of sub-populations. However it has always been a difficult task due to the fact that bacteria are small particles (usually 1–3 μm). For such small particles, light scatter signal is on the threshold level, and many positive events may be lost. Here we will present the development of a procedure for sorting of the gram negative bacteria *Y. pestis* from environment samples. We will show that the application of nano-magnetic particles, as a tool for the immunomagnetic labeling and separation of the bacteria, enables fast sorting in high and low bacterial concentration down to 10^5 cfu/ml. The nano-metric physical size of the immunospecific labeling particles disguises them from the FACS detectors; hence the bacterial population becomes the major population as opposed to being "rare events population" when using standard micro-magnetic beads for pre-enrichment.

The procedure of separation and collection of bacteria enables sensitive detection and characterization methods of bacteria from complex samples.

E. Zahavy (✉) • H. Abramovich • E. Freeman • S. Yitzhaki
Department of Infectious Diseases, Israel Institute for Biological Research, P.O. Box 19, Ness-Ziona 74100, Israel
e-mail: eranz@iibr.gov.il

R. Ber • D. Gur • S. Maoz
Department of Biochemistry and Molecular Genetics, Israel Institute for Biological Research, P.O. Box 19, Ness-Ziona 74100, Israel

E. Zahavy et al. (eds.), *Nano-Biotechnology for Biomedical and Diagnostic Research*, Advances in Experimental Medicine and Biology 733, DOI 10.1007/978-94-007-2555-3_3,

Keywords Fluorescent nano-particles • Magnetic nano-particles • Pathogenic bacteria • Flow cytometry • FACS

3.1 Introduction

Among the most deadly pathogens found in group A of the bio-terror list are *Bacillus anthracis* and *Yersinia pestis*. In case of deliberate contamination as a result of a bioterror attack a rapid and a reliable detection is crucial for isolating of the incident area, treating it and protecting the remaining un-exposed population. In such scenarios the detection teams which are required to collect and identify the potential hazards will face complex samples, sometimes from unknown origin, and many times these samples will contain endemic microrganisms, spores or inorganic debris. For such requirements, one needs to develop tools for detection of the harmful pathogens. Those methods should be fast, sensitive, highly credible, suitable for simultaneous agent detection (multiplex analysis) and applicable in complex environmental samples. In this manuscript we shall focus on Flow Cytometry (FCM) as a diagnostic and sorting tool for bacteria based on specific antibodies which are fluorescently labeled with fluorescent semiconductor nano-crystals (q-dots). It will be shown that such combination can increase the diagnostic specificity and enables multiplex analysis of two different bacteria in single measurement. Moreover, we will show in here a novel way of collecting the bacterial population from any samples using the combination of nano-immunomagnetic separation and flow-cytometry preparative sorting. This new approach will improve both reliability and efficiency of the downstream detection process of pathogens bacteria, which is a crucial point in the counterterrorism actions.

Immunofluorescence detection of bacteria by FCM have been described in the past (Zahavy et al. 2003; Yitzhaki et al. 2004; Nebe-von-Caron et al. 2000; Shapiro 2000; Stopa 2000; Falcioni et al. 2006; McHugh and Tucker 2007). Recently other works (Bartek et al. 2008; Venkatapathi et al. 2008) have shown the identification of bacteria by light scatter parameters solely. However such criteria suffer from false positive results. Hence, the use of specific fluorophore labels, such as conjugated antibodies with light scatter parameters is still a crucial combination for reliable analysis by FCM. The spectroscopic properties of organic fluorophores are characterized by narrow excitation band and broad emission band. Hence, such fluorophores require multi-wavelength excitation in order to observe several emission lines from different fluorophores, and this limits their use as simultaneous labels. As opposed to the organic fluorophores, q-dots are characterized by broad excitation band and narrow emission band. Such optical setting enables single excitation source that simultaneously can excite several q-dots with distinctive emission bands, which can easily be separated by optical filters. During the last decade fluorescent q-dots have been chemically modified on their surface for the introduction in biological matrices (Gerion et al. 2001; Goldman et al. 2002; Michalet et al. 2005). The potential of using q-dots as fluorescent labels for biological systems was studied by several research groups which are aiming for developing new tool for cell studying (Bruchez et al. 1998), cell staining (Yitzhaki et al. 2005), fluoro-immunoassays (Goldman et al. 2002), pathogen immunodetection (Hahn et al. 2008; Liu et al. 2007; Su and Li 2004), *in-vivo* imaging (Michalet et al. 2005), immunophenotyping by flow cytometry analysis (Chattopadhyay et al. 2006; Summers et al. 2010; Tarnok 2010; Zahavy et al. 2010; Godfrey et al. 2009; Jaron and Godfrey 2009), FRET biological sensor (Geisler et al. 2010; Medintz and Mattoussi 2009), and multiplexed analysis for DNA analysis (Hahn et al. 2001). In here we will show that the use of the fluorescent q-dots can contribute to current immunofluorescence analysis of biological hazards in two ways: (1) Different conjugates can be used for multi-parameter analysis by FCM in order to achieve fast, sensitive and accurate analysis of a single biological threat, such as *Bacillus anthracis* (*B. anthracis*) spores. (2) Series of fluorescent q-dots that are attached to different antibodies can be utilized for the development of multiplex analysis of two (or more) biological hazards in single measurements.

Apart from analysis of the bacteria, preparative sorting of bacteria by FACS has a tremendous advantage for enhancing sensitivity and selectivity of any downstream detection and analysis methods. In this work we have studied bacterial sorting of the gram negative bacteria *Yersinia pestis* (*Y. pestis*) from implanted environment samples. Such procedure of separation and collection of bacteria from complex environments can enable sensitive detection and characterization of the bacteria. Sorting bacteria by FACS requires significant light scatter parameter in addition to the fluorescence labeling. This is essential for threshold criteria and for analysis process. Lacking of good light scattering analysis could lead to overflow of background events that might mask the target bacterial events (Nebe-von-Caron and Muller 2010; Shapiro 2000, 2003). This would be even more crucial when enrichment steps are not included, hence other bacteria in the samples can dominate the slow growing bacteria, *Y. pestis*. In order to overcome such problems, an IMS (ImmunoMagnetic Separation) process for pre enrichment of bacteria was used. Although the use of micro-magnetic beads for pre-enrichment of bacteria is well established (Fisher et al. 2009; Jenikova et al. 2000; Stevens and Jaykus 2004; Zhao et al. 2009) in here we would show the benefit of using nano-magnetic beads as pre enrichment of bacteria prior to the FACS sorting. In previous work (Fuchslin et al. 2010) it was shown that the use of pre enrichment of the pathogen *Legionella pneumophila* improved its detection limits by flow cytometry. In here we will present how pre-enrichment of *Y. pestis* by nano magnetic beads is superior to the pre-enrichment by microbeads. Moreover, such pre-enrichment enabled us to perform fast and efficient bacterial sorting, and achieve clean bacterial "preparate" overcoming the long step of isolation by the traditional agar plate cultivation.

3.2 Results

3.2.1 Fluorescent Immunolabeling and FCM Detection of Pathogenic Bacteria with Fluorescent Nano-Particles (q-dots)

3.2.1.1 Flow Cytometry Specific Analysis of *B. anthracis* Spores by q-dots Conjugates Staining

Specific analysis of bacterial cells by flow cytometry required initial light scatter gating prior to the immunostaining by the specific Q-dots conjugates. The contribution of the light scatter parameters to the *B. anthracis* spore analysis has been described previously (Zahavy et al. 2010). The light scatter parameters are not sufficient to distinct between different types of spores, however they are crucial for the initial gating and for preliminary filtration of the desired population. Overall, light scatter parameters can reduce undesired population by 10–40% by implying the correct gating as the primary gate selection for spores. The major and crucial selectivity arises from the immunofluorescence labeling. For single and double immunospecific identification of the target spores, we have prepared two q-dots conjugates: q-dots585-IgGα*B. anthracis* and q-dots655-IgGα*B. anthracis*. Both conjugates were characterized (Zahavy et al. 2010) and had maintained the immunospecific properties of the antibodies and the fluorescence properties of the q-dots. It was shown by flow cytometry analysis that q-dots585 can be exclusively detected on FL2 channel where q-dots655 is observed on the FL3 and FL4 channels. In the flow cytometry optical set up, both FL2 and FL3 channels are related to the 488 laser excitation where their emission lines are 585 nm and 670 nm respectively. The FL4 channel is related to the 633 nm laser excitation with an emission line of 660 nm. Hence it is quite obvious to see q-dots655 emission both in FL3 and FL4 detectors. However, since bacterial cells are small, and

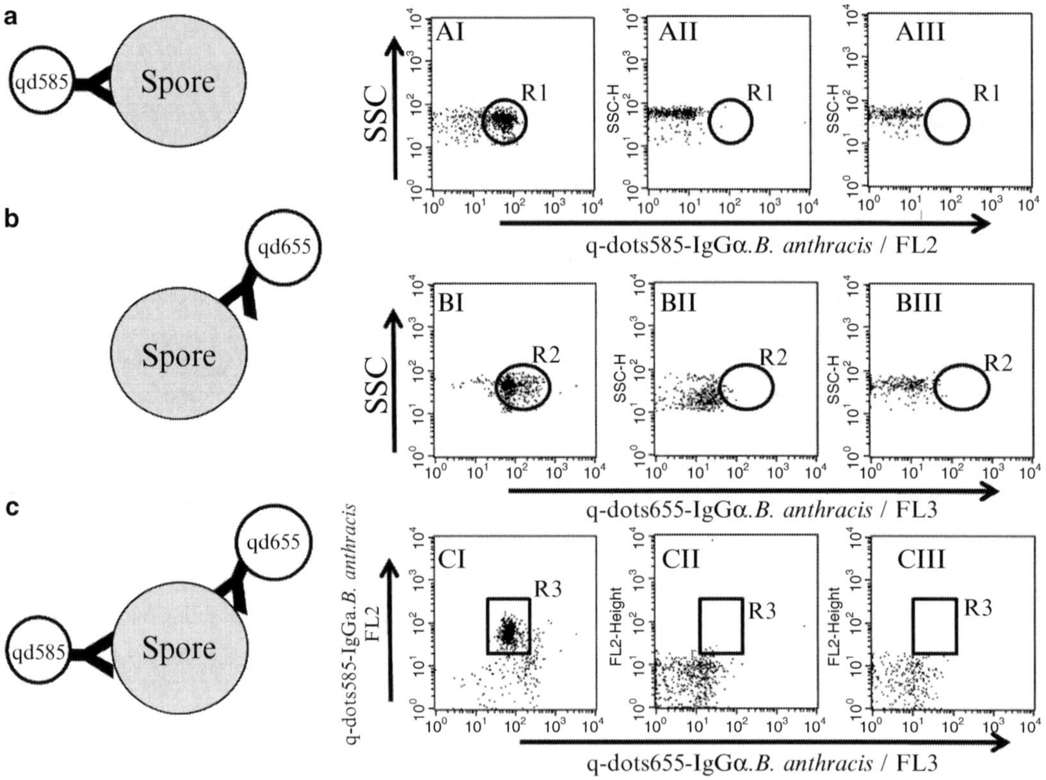

Fig. 3.1 Schematic representation and the FCM dot plots of: (**a**) Single labeled spore with q-dots585-IgGα*B. anthracis*. (**b**) Single labeled spore with q-dots655-IgGα*B. anthracis*. (**c**) Double labeled spore with q-dots585-IgGα*B. anthracis* and q-dots655-IgGα*B. anthracis*. The labels I, II and III refer to samples of *B. anthracis*, *B. thuringiensis I.* and *B. subtillis* spores respectively

might have weak light scatter and fluorescence signals, we used only the channel data related to the 488 nm excitation: FL2 and FL3. Overall we have used light scatter gating and the FL2/FL3 gating in order to form a selective region for the detected spores. Evaluation of the immunospecific staining was performed by analyzing samples of the stained *B. anthracis* spores in comparison to relative phylogenetically close spores such as *B. thuringiesis I.* (BTI) or *B. subtillis* (B. sub). In Fig. 3.1, one can see the resulting specific region, R1 or R2, of the stained spores for single labeled spore stained with q-dots585-IgGα·*B. anthracis*, or q-dots655-IgGα·*B. anthracis*. The region parameters are SSC/FL3 or SSC/FL2 and they are shown in Fig. 3.1a, b respectively. For the double staining, the region parameters are FL2/FL3 as shown in Fig. 3.1c as R3. In *B. anthracis* spore samples the regions contain more than 90% of the events as positive recognized spores, while using BTI or B. sub spore samples, the positive events are highly decreased. This implies that the labeling is indeed specific with minimal false-positive events. The specificity factor has been quantified by measuring the percentage of events in the designated spore region (R2) of the competitive spores in comparison to the *B. anthracis* spores dot-plots, normalized to 10,000 events. From the gating process we can see that in the specific region for the stained *B. anthracis* spores, the percentage of *B. subtilis* is ca ~1% and higher for *B. thuriengensis I.* which is 1–3%. This might be due to the high genetic resemblance between *B. thuringiensis I.* and *B. anthracis*. However, using the double labeling, we achieved a decrease of the background events to 0.01%.

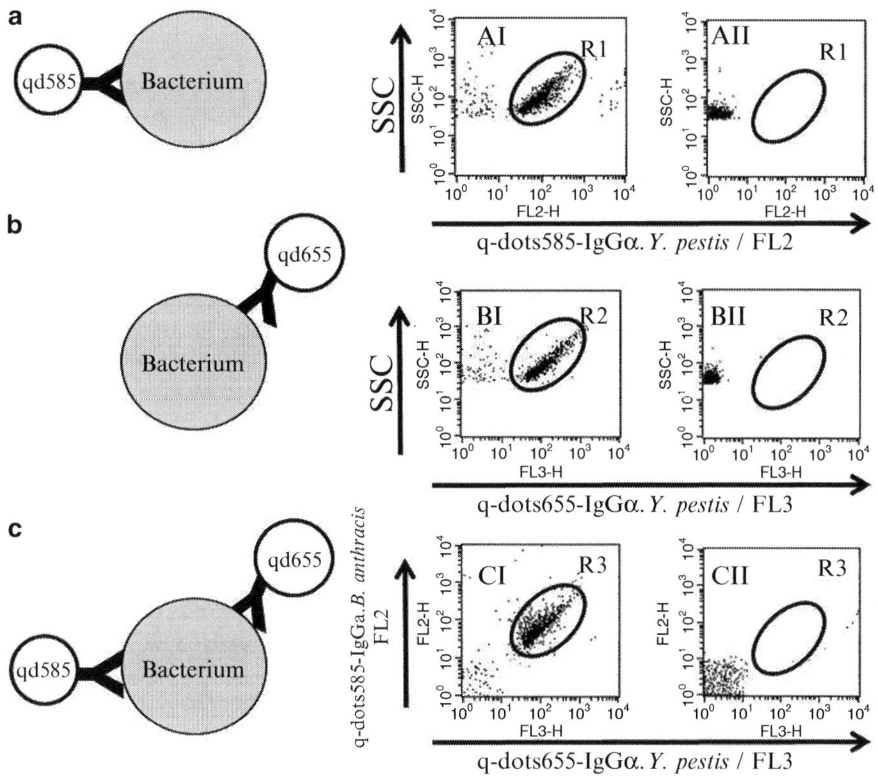

Fig. 3.2 Schematic representation and the FCM dot plots of: (**a**) Single labeled sample with q-dots585-IgGα*Y. pestis*. (**b**) Single labeled sample with q-dots655-IgGα*Y. pestis*. (**c**) Double labeled sample with q-dots585-IgGα*Y. pestis* and q-dots655-IgGα*Y. pestis*. The labels I and II refer to samples of *Y. pestis* and *Y. enterecolitica* respectively

3.2.1.2 Flow Cytometry Specific Analysis of *Y. pestis* Bacteria by q-dots Conjugates Staining

The same methodology was implemented for the detection of the pathogenic bacterium *Y. pestis*, with two new conjugates: q-dots585-IgGα*Y. pestis* and q-dots655-IgGα*Y. pestis*. The new prepared conjugates were found specific toward *Y. pestis* vs. variety of gram negative bacteria such as *Y. pseudotuberculosis, E. coli* (TG1), *S. aureus*, *E. faecalis*, *Y. enterocolitica* and *P. aeruginosa*. The selection regions, R1 (SSC/FL2) and R2 (SSC/FL3), for single-labeling bacterial samples with q-dots585-IgGα*Y. pestis* or q-dots655-IgGα*Y. pestis* are shown in Fig. 3.2a, b, respectively. Results of double labeling of the bacteria in FL2/FL3 region (R3) are shown in Fig. 3.2c. Quantification of the specificity factor has been evaluated as described for the spores analysis above. From the gating process we can see that in the specific region for the stained *Y. pestis*, the percentage of control bacteria is ca ~0.1–1% with q-dots585-IgGα*Y. pestis* or q-dots655-IgGα*Y. pestis* single labeling with most of the bacteria. Samples of *Y. pseudotuberculosis*, which is phylogenetically related to the *Y. pestis*, results in higher cross reactivity of ca ~2% or 10% with q-dots585-IgGα*Y. pestis* or q-dots655-IgGα*Y. pestis* labeling respectively. Double labeling of the samples reduce the cross reactivity to 0–0.05% of false events, and 0.3% with the *Y. pseudotuberculosis* samples.

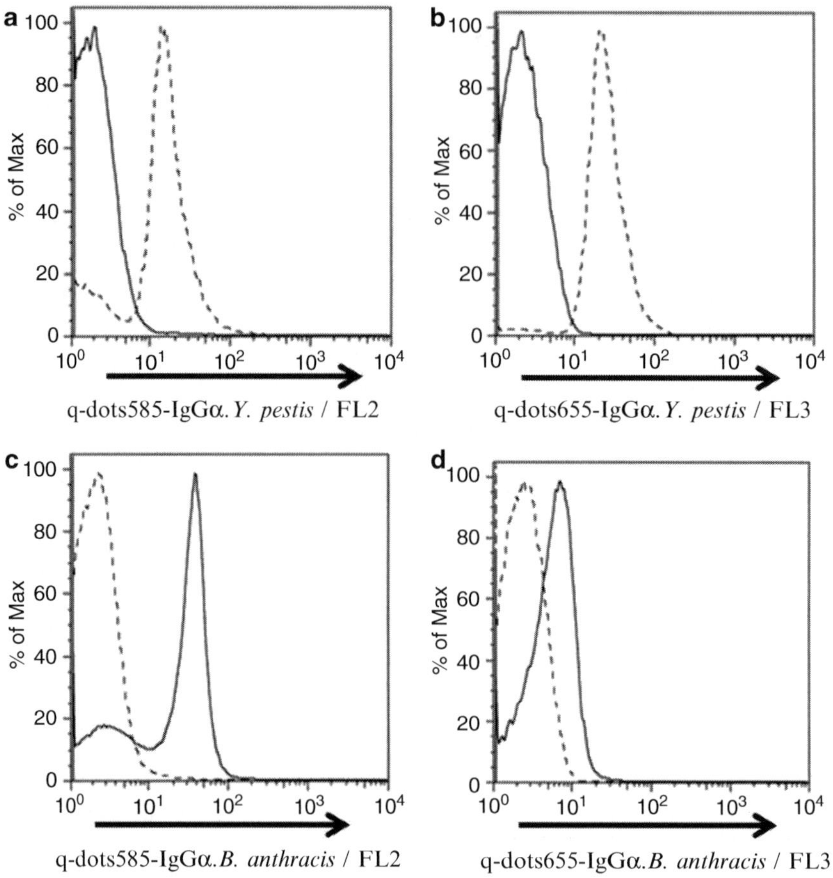

Fig. 3.3 Flow Cytometry fluorescent histograms of *Y. pestis* (*dashed line*) and *B. anthracis* spores (*solid line*). (**a**) FL2 histograms after staining with q-dots585-α*Y. pestis*. (**b**) FL4 histograms after staining with q-dots655-α*Y. pestis*. (**c**) FL2 histograms after staining with q-dots585-α*B. anthracis*. (**d**) FL4 histograms after staining with q-dots655-α*B. anthracis*

3.2.1.3 Application of the q-dots Conjugates for Multiplex Analysis of *Y. pestis* and *B. anthracis* Spores

As seen for both *B. anthracis* spores and the *Y. pestis* bacteria, both single labeling resulted in high specificity values, whereas by using the double labeling, specificity is increased at least by one log. However, in order to achieve the immediate FCM diagnosis of *B. anthracis* spores or *Y. pestis* bacteria in a sample in one measurement we need to prepare a mixture of single specific conjugates for each pathogen, with different fluorescence labeling. Since the main obstacle in achieving simultaneous detection is attributed to the possible cross reaction between the antibodies and the different pathogens, we have measured the cross reactivity of the prepared q-dots-Antibody conjugates with the two pathogens, as shown in Fig. 3.3. The FCM analysis resulted in the fluorescence histograms of the different bacteria species, *B. anthracis* and *Y. pestis* in solid and dashed lines respectively, and with different staining. In Fig. 3.3a, the pathogens samples were stained with q-dots585-IgGα*Y. pestis*, and accordingly were detected by the FL2 channel in the FCM. In Fig. 3.3b the samples were stained with q-dots655-IgGα*Y. pestis*, and accordingly were detected by the FL3 channel in the FCM. It can be seen that both conjugates facilitate selective staining of the *Y. pestis*. In fact, by measuring the FL2 or FL3 mean values of the *Y. pestis* to the *B. anthracis*, we obtained a ratio of 10,

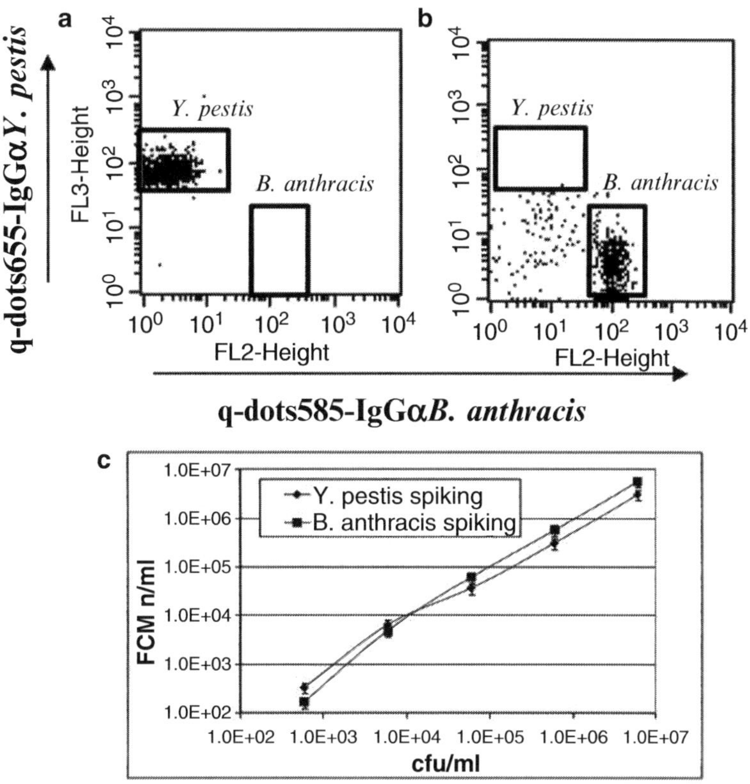

Fig. 3.4 Dot plots of *Y. pestis* (**a**) and *B. anthracis* spore (**b**) samples in presence of mix conjugates for simultaneous detection, q-dots655-IgGα*Y. pestis* and q-dots585-IgGα*B. anthracis*. (**c**) Correlation graph between the bacteria concentration in the samples to the FCM analysis and counting

which reflects a specific staining and good terms for further simultaneously staining. Figure 3.3c, 3.3d refers to samples stained with q-dots585-IgGα*B. anthracis* and with q-dots655-IgGα*B. anthracis* respectively. From the data shown in Fig. 3.3c a selectivity factor of ca ~ 18 is derived, which indicates high specific staining by the q-dots585-IgGα*B. anthracis* conjugate. However, Fig. 3.3d shows poor selectivity staining of the q-dots655-IgGα*B. anthracis* conjugate, with selectivity factor of only 3 (less than one "decade"). Following these results we decided to use the conjugates: q-dots585-IgGα*B. anthracis* and q-dots655-IgGα*Y. pestis* for multiplex analysis of both pathogens. These conjugates are selective, have no cross reaction between the two pathogens and their labeling intensity is one magnitude higher than the background.

The multiplex diagnosis of both pathogens require the assembly of both pathogen populations into one dot plot, where each population is in a distinctive gate, as described in Fig. 3.4. The selective analysis of the different bacteria species is possible due to the specific fluorescent conjugates that we have prepared and carefully examined. The fluorescent dot plot consists of the specific channel (FL2) for the *B. anthracis* based on the conjugate q-dots585-IgGα*B. anthracis*, and a specific channel (FL3) for the *Y. pestis* based on the conjugate, q-dots655-IgGα*Y. pestis*. Such configuration enables us to separately identify the *B. anthracis* spores on the FL2 channel and *Y. pestis* on the FL3 channel. The fluorescence dot-plots shown in Fig. 3.4, represent a series of experiments where in each sample we introduced either one of the two pathogens in different concentrations. In all samples both conjugates, q-dots655-IgGα*Y. pestis* and q-dots585-IgGα*B. anthracis,* were present. It is clearly shown by the dot-plots that each bacteria is lightning the expected channel for its identification.

We can see the events related to *Y. pestis* filling the lower right quadrant on the channel FL3 and the events related to the *B. anthracis* spores filling the upper left quadrant on the channel FL2. Moreover, while using mixed samples of both pathogens we have revealed both populations distinct in their designated FL2/FL3 gates (data not shown). The number of events related to each bacterium is correlated to the concentration of the samples. From the number of events for each bacteria (gate), the flow rate (v, μl/min) and the time of the sampling (t, min), we can calculate the bacterial concentration in the samples. The concentration in relation to the events is shown in Fig. 3.4c. The graph shows high correlation between the actual pathogen concentrations in the samples and the concentration calculated from the events number as it appear in the FCM analysis. This correlation is firm from 10^3 up to 10^7 cfu/ml for both pathogens.

3.2.2 Preparative Sorting of *Y. pestis* Bacteria from Complex Samples

Beyond the flow cytometry diagnosis of bacteria in complex samples, it would be a great advantage to apply preparative sorting on the bacterial population. Such ability, in particular from complex samples, would allow better down-stream diagnostic steps in term of sensitivity, selectivity, biological viability and more. As described in the introduction, complex samples could be un-expected and not as we know them now, such as swab samples from envelops, office's desk, pave walks etc., hence pulling out the pathogen bacteria could be some times crucial for detection. In here we present sorting of the pathogen *Y. pestis* in a model of complex samples which are swabs collection from local dirt, spiked with the bacteria. In our experiment, the key to successful sorting is given by reducing the un-wanted micro-organisms which co-exist in the samples along with our spiked bacteria. The major tool to observe the ratio between our spiked bacteria to the rest of the microorganism is by plating the samples, before and after the sorting, on rich agar plates. Since *Y. pestis* is a slow growing bacteria (48 h for visible colonies on agar plates) and any local bacteria or fungi will grow faster and will easily dominate the agar plate, it would be easy to observe an un-pure sorting by such plating. This will enable us very sensitive tool to learn about our sorting purity and efficiency and to ensure that our new method is valid.

3.2.2.1 Sorting Limitation

As claimed in the introduction, the diagnostic and sorting of bacterial species by flow cytometry is a challengeable task since bacteria are small entities with weak light scatter signals, which sometime are on the edge of the background or debris signals. Light scatter parameters are essential for threshold criteria in the sorting and for the analysis process and lacking of good light scatter analysis could lead to overflow of background events that would disguise the target bacterial events. Beyond light scatter analysis, it is also crucial that within the complex samples the target bacteria won't become a rare population which is difficult to sort to high purity as explained in here. Since sorting is matter of cells or bacteria in droplets and collecting the correct drops, the distribution of the cells within the drops is an important issue to achieve pure and efficient sorting. One would like to spread the cells within the drops not only one cell per drop, but also with empty drops between occupied drops. This is because sorting in high purity mode is performed not on single drop basis but on the basis of drops "envelope" of two to three drops. Hence, gathering of different cells in neighboring drops will be considered as "coincidence events" (Shapiro 2003). Coincidence events, which include the desired sorted bacteria but also a foreign one in the same "envelope", will be collected as positive and will cause the final preparation to be impure. In order to reduce to minimum the coincidence events rate we have to control the factors affecting it by adjusting the sorting

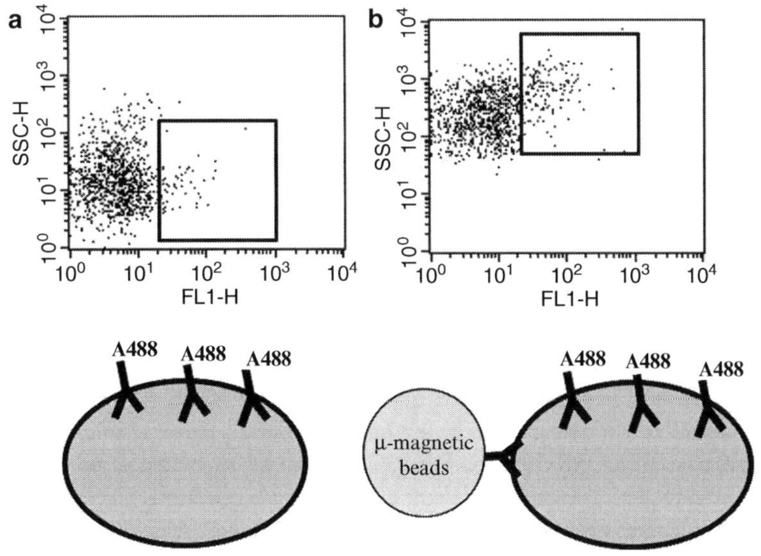

Fig. 3.5 FACS dot plot and schematic presentation (*bottom*) of (**a**) IgG-Alexa488 labeled *Y. pestis* bacteria in complex sample. (**b**) IgG-micro magnetic beads and IgG-Alexa488 labeled *Y. pestis* bacteria in complex sample

parameters accordingly. The rate of coincidence event, P, is described by the Poisson distribution (Eq. 3.1), where f is the ratio between positive events and the total events, u is the total events rate and DDF is the drops rate. The equation shows that the coincidence events rate, P, increases when there are too many events per drops and the sorted population is rare within the total events. P will decrease when the ratio f increase and the DDF will increase. For example, we have calculated (not shown here) that when the drops rate is ten times higher than the total events rate (DDF >10) and the positive population is more than 50% ($f > 0.5$), the chances of coincidence events decrease to less than 0.001%.

$$\mathrm{P} = e^{-2f(DDF|u)} \tag{3.1}$$

However, dealing with suspected pathogen-infected complex samples, one has to face an unknown type and quantity of debris (organic and inorganic) or different microorganisms. As a result, we have seen that in such samples our spiked population, in concentration of 10^5–10^8 cfu/ml, the pathogenic bacteria are appearing as rare events which do not exceed population ratio of more than 1% in most cases. In the dot plots presented in Fig. 3.5a, it is shown that the labeled bacterial population in such a sample is less than 1%, which is considered to be as rare population; hence sorting will suffer from high rate of coincidence events. To increase the target bacterial population in the sample we have adopted an immunomagnetic separation (IMS) of the targeted bacteria. Commercial Streptavidin-micro-magnetic beads (from Dynal, 1 µm) were coated with biotinilated IgG α *Y. pestis*, to achieve specific separation by applying an external magnetic field and washing steps. However, in order to achieve reasonable collection yields (>40%) it is necessary to maintain a 100-fold excess of the beads concentration to the bacteria. Moreover, the micro-magnetic beads are observed in the light scatter parameters of the flow cytometry analysis in the same parameters as the bacteria. This limitation of the micro magnetic beads causes the bacteria population to remain as rare events during the sorting, as can be seen in Fig. 3.5b. In bacterial concentration range of 10^6–10^8 cfu/ml we found out that after micro-magnetic IMS procedure the ration of positive bacterial events is ca ~ 1–10%. Although we have eliminated most of the environmental noise by the IMS, the magnetic beads themselves become the main contribution of the noise. In such condition of rare population sorting, the probability of coincidence events is increases. Hence we have to keep the ratio of the drops number in the sorter to the total events on at least 10 to 1, which can maintain low probability ($P<0.1\%$) of coincidence events. In such condition, sorting in high pressure and 60,000 drops per second we are achieving sorting rate of ca ~ 100 bacteria per second, hence the minimal time to collect one million bacteria is ~2 h.

3.2.2.2 Application of Nano-Magnetic Beads for the Pre-Enrichment

In order to combine the advantages of the magnetic pre-enrichment process and still to avoid the disadvantages that arise in the sorting process, it is necessary to eliminate the micro-magnetic signals in the flow-cytometry detectors. For this purpose, coated magnetic particles with low light scatter signals are needed. Since the initial gating on the events analyzed in flow-cytometry detectors are light scatter signals (FSC/SSC), once the light scatter of the magnetic particles would be under the threshold it would be possible to eliminate any excess of particles from the detectors. It is also important to maintain the light scatter signals of the detected bacteria above the threshold. Such configuration would increase the bacterial population over debris, particles etc. For this we have measured 3 different diameters of coated magnetic beads: 50, 100 and 200 nm. All three types are carboxyl coated beads (Chemicell, nano-screenMAG-CMX) and they were covalently attached to IgG α *Y. pestis*, by EDC coupling (Hermanson 1996), for specific immunoenrichment. The 100 and 200 nm beads contain multiple magnetic dipoles; hence they require only simple magnetic separation. The 50 nm beads contain single magnetic dipole, and hence they require strong magnetic separation such as the application of the magnetic columns by Miltenyi. Flow cytometry analysis of the different beads revealed that in both 100 and 200 nm diameter beads, light scatter signals are still high and cannot be excluded by threshold conditions, without losing the bacterial population. Using smaller size particles such as the 50 nm beads, it was shown that the light scatter signals are weak enough to be excluded in the flow cytometry light scatter detectors. Hence, free nano-magnetic particles in 50 nm diameter, which are not attached to any bacteria, are excluded both in light scatter and fluorescence parameters (data not shown). This was also confirmed by scanning electron microscope imaging of labeled bacteria with the nano-magnetic-antibodies conjugate. In the SEM imaging, Fig. 3.6d, the nano-magnetic particles appears as white dots, and can be seen attached to the bacterial surface. Their size is ca ~1e6 smaller as compared to the bacteria, hence their light scatter signals are negligible compared to the bacteria's.

Overall, for the collection of *Y. pestis* bacteria from complex samples, we have adapted the two step separation. The first step includes the IMS, using the nano-magnetic IgG conjugate on the Miltenyi separation column, which take less than an hour. The second step includes immunofluorescence labeling of the sample after the IMS, and direct sorting on the FACS (BD, FACSVantage). Application of this method for sorting *Y. pestis* bacteria from complex environment can be seen in the relevant flow cytometry dot plots in Fig. 3.6. Figure 3.6a represent analysis of a complex field sample after the immunomagnetic separation (IMS). It is clear that after the IMS, the sample still contain many of the debris that had been passing thru the IMS. This can be observed in the light scatter parameter dot plots, where region R1 is marked. However, in the designated region for the fluorescence immunolabeled bacteria, marked as R1, there are less than 0.01% false events. The same samples, spiked with 10^6/ml or 10^7/ml bacteria show different results, Fig. 3.6b, c respectively. Now in the fluorescence dot plots we can see 10% and 70% out of the total population as positive bacteria. Calculating the possible sorting speed, with the desired limitation of reducing coincidence events to $P<0.01\%$, with population of $f=60$–80%, DDF = 60,000 and u = DDF/10, we can estimate that collection of million bacteria will take 5–10 min.

The combined nano-IMS and FACS sorting was demonstrated on several samples spiked with *Y. pestis* bacteria in the range of 10^5–10^7 cfu/ml into samples of *B. anthracis* spores in known

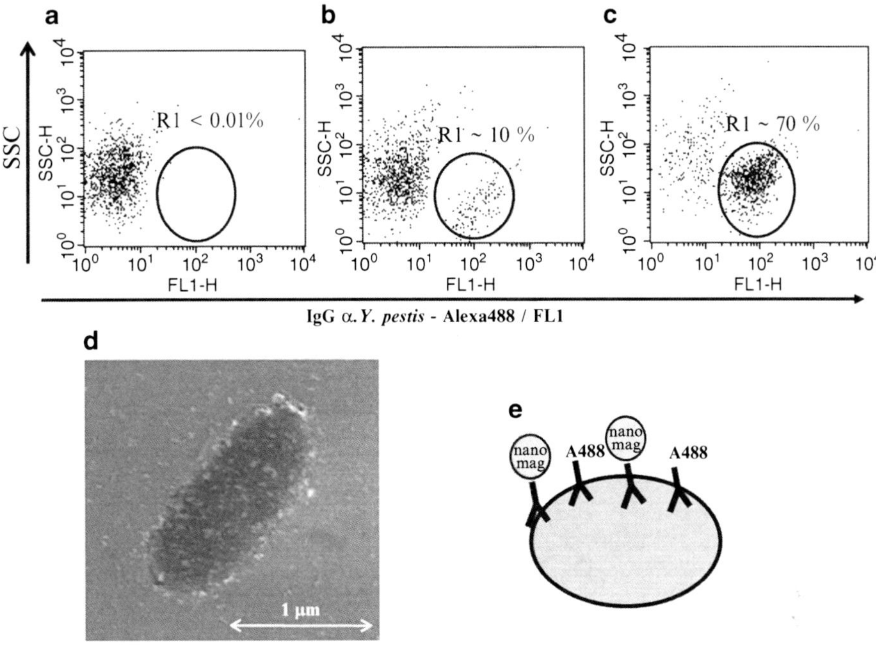

Fig. 3.6 (**a–c**) Dot plots of complex samples spiked with 0, 1×10^6 and 1×10^7 cfu/ml *Y. pestis* bacteria, after nano-magnetic separation and fluorescence labeling of the samples with IgG α *Y. pestis* – Alexa488. (**d**) SEM imaging of the nano-magnetic beads (*white dots*) on the bacterial surface. (**e**) Schematic presentation of the labeled bacteria

Table 3.1 Results of pre-enrichment by nano-IMS and FACS sorting of *Y. pestis* from spiked samples

Spiking *Y. pestis*	Impurity in the sample	Initial mixture ratio	Mixture ratio after nano-IMS	Mixture ration after FACS	Yield	Collection time to million bacteria (min)
10^6 cfu/ml spiked *Y. pestis*	*B. anthracis* 10^5 cfu/ml	10:1	10^4:1	10^5:1	50	~60
	B. anthracis 10^4 cfu/ml	100:1	10^4:1	10^5:1	60	~60
	Field	10:1	1000:1	10^5:1	30	~60
	Field	100:1	10^4:1	10^5:1	30	~60
10^7 cfu/ml spiked *Y. pestis*	*B. anthracis* 10^5 cfu/ml	100:1	100:1	10^5:1	70	~10
	B. anthracis 10^4 cfu/ml	1000:1	10^4:1	10^5:1	80	~10
	Field	100:1	1000:1	10^5:1	40	~10
	Field	1000:1	10^5:1	10^6:1	40	~10

concentration and in field samples where debris and microorganism are not known. For the field samples, swabs spread from local pave ways have been used. Estimation of the microorganism counts in the field samples is estimated by plating on rich agar plates. The results of the IMS and FACS sorting of different spiked *Y. pestis* samples is summarized in Table 3.1. As can be seen, while the concentration of *Y. pestis* is in the range of 10^5–10^6 cfu/ml and the samples are either mixed sample with fixed concentration of *B. anthracis* spores or field samples, we were able to collect target bacteria in a rate of million per 60 min. Moreover, from mixture ratios where impurities are 1–10% we have reached mixture of less than 1 impurity in 100,000 *Y. pestis* bacteria. Such

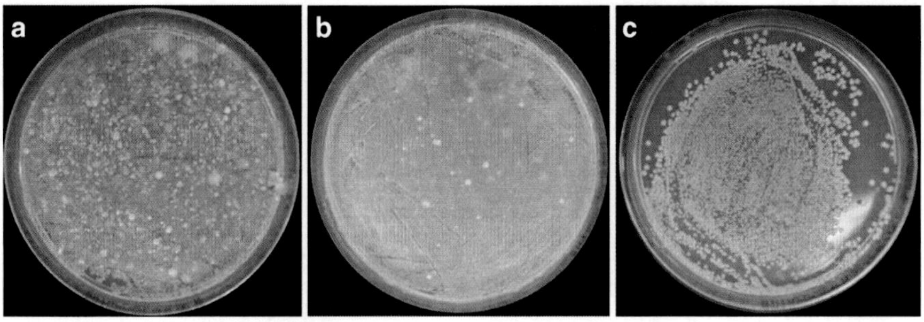

Fig. 3.7 Plate counting of field samples spiked with 10^7 cfu/ml *Y. pestis* bacteria: (**a**) before any separation. (**b**) After nano-magnetic separation (nano-IMS) process. (**c**) After the FACS sorting

results couldn't be achieved using samples without any pre enrichment by IMS nor couldn't be achieved with IMS pre enrichment based on micro magnetic beads. Using samples concentration of 10^7 cfu/ml or more of the spiked bacteria, we can collect million bacteria within 10 min. Moreover, in all of the combined IMS/FACS separations, we have ended with pure bacteria samples, where impurities were excluded by 5–6 logs. As shown in Fig. 3.7, agar plating of the samples after IMS show that the preparation still contain ca 1:100 to 1:1000 impurities of microorganism. After the FACS sorting step, it is hard to detect any residual microorganism beside the bacterial species that was sorted.

3.3 Conclusion

This paper summarizes two major contributions of nano-materials to microbiological world in flow cytometry. The first one includes the application of the unique spectral characterization of fluorescent nano-crystals (q-dots) to multiplex diagnostic of bacteria in flow cytometry. We have shown that by applying these unique fluorescent labels one can improve the selectivity of bacterial diagnostic tools, such as the bio-terror candidates *B. anthracis* or *Y. pestis*. Moreover, we have shown that multiplex diagnosis of both pathogenic bacteria is possible. The q-dots that had been used are q-dots585 and q-dots655, both excited by the 488 laser and are separately observed on the FL2 / FL3 channels respectively.

For sorting purposes of the bacteria we found that application of nano-magnetic beads is necessary for pure and fast sorting from complex samples. The pre-enrichment of the bacteria from the samples enable the bacterial sorting within the FACS, after immunofluorescence labeling. Using nano-magnetic particles (50 nm) for the pre enrichment is of great advantage, since the nanometric physical size disguises them from the FACS detectors; hence the bacterial population becomes the major population and not "rare events population" as by using standard micro magnetic beads for enrichment. This development enables a rapid and accurate sorting procedure by the FACS. We have shown here sorting of ca ~ million bacteria in 5–10 min, and overall reduction of contaminating bacteria by 5–6 orders of magnitude. Such procedure of separation and collection of bacteria, enables sensitive detection and characterization methods of bacteria from complex environments.

References

Bartek, R., Venkatapathi, M., Ragheb, K., Banada, P. P., Hirleman, E. D., Larry, T., & Robinson, J. P. (2008). Automated classification of bacterial particles in flow cytometry by multiangle scatter measurement and support vector machine classifier. *Cytometry, 73A*, 369–379.

Bruchez, M., Jr., Moronne, M., Gin, P., Weiss, S., & Alivisatos, A. P. (1998). Semiconductor nanocrystals as fluorescent biological labels. *Science, 281*, 2013–2016.

Chattopadhyay, P. K., Price, D. A., Hatper, T. F., Betts, M. R., Yu, J., Gostick, E., Perfetto, S. P., Goepfert, P., Koup, R. A., de Rosa, S. C., Bruchez, M. P., & Roederer, M. (2006). Quantum dot semiconductor nanocrystals for immunophenotyping by polychromatic flow cytometry. *Nature Medicine, 12*, 972–977.

Falcioni, T., Manti, A., Boi, P., Canonico, B., Balsamo, M., & Papa, S. (2006). Comparison of disruption procedures for enumeration of activated sludge FLoc bacteria by flow cytomery. *Clinical Cytometry, 70B*, 149–153.

Fisher, M., Atiya-Nasagi, Y., Simon, I., Gordin, M., Mechaly, A., & Yitzhaki, S. (2009). A combined immunomagnetic separation and lateral flow method for a sensitive on-site detection of bacillus anthracis spores - assesment on water and dairy products. *Letters in Applied Microbiology, 48*(4), 413–418.

Fuchslin, H. P., Kotzsch, S., Keserue, H.-A., & Egli, T. (2010). Rapid and quantitative detection of legionella pneumophila applying immunomagnetic separation and flow-cytometry. *Cytometry, 77A*, 264–274.

Geisler, D., Charbonniere, L. J., Ziessel, R. F., Butlin, N. G., Lohmannsroben, H.-G., & Hildebrandt, N. (2010). Quantum dot biosensors for ultrasensitive multiplexed diagnostics. *Angewandte Chemie, International Edition, 49*, 1396–1401.

Gerion, D., Pinaud, F., Williams, S. C., Parak, W. J., Zanchet, D., Weiss, S., & Alivisatos, A. P. (2001). Synthesis and properties of biocompatible water-soluble silica-coated CdSe/ZnS semiconductor quantum dots. *The Journal of Physical Chemistry B, 2001*, 8861–8871.

Godfrey, W. L., Zhang, Y. Z., Jaron, S., & Buller, G. M. (2009). Qdot nanocrystal conjugates in multispectral cytometry. *The Journal of Immunology*, 182, 42.12.

Goldman, E. R., Anderson, G. P., Tran, P. T., Mattoussi, H., Charles, P. T., & Mauro, J. M. (2002). Conjugation of luminescent quantum dots with antibodies using an engineered adaptor protein to provide new reagents for fluoroimmunoassays. *Analytical Chemistry, 74*, 841–847.

Hahn, M., Gao, X., Su, J. S., & Nie, S. (2001). Quantum-dot-tagged microbead for multiplexed optical coding of biomolecules. *Nature Biotechnology, 19*, 631–635.

Hahn, M., Keng, P. C., & Krauss, T. D. (2008). Flow cytometric analysis to detect pathogens in bacterial cell mixtures using semiconductor quantum dots. *Analytical Chemistry, 80*, 854–872.

Hermanson, G. H. (1996). *Bioconjugate techniques*. San Diego: Academic.

Jaron, S., & Godfrey, W. L. (2009). Multicolor flow cytometry using only Qdot conjugates primary antibodies. *The Journal of Immunology*, 182, 42.12.

Jenikova, G., Pazlarova, J., & Demnerova, K. (2000). Detection of salmonella in food samples by the combination of immunomagnetic separation and PCR assay. *International Microbiology, 3*, 225–229.

Liu, Y., Brandon, R., Cate, M., Peng, X., Stony, R., & Johnson, M. B. (2007). Detection of pathogen using luminescent CdSe/ZnS Dendron nanocrystals and porous membrane immunofilter. *Analytical Chemistry, 79*, 8796–8802.

McHugh, I. O. L., & Tucker, A. L. (2007). Flow cytometry for the rapid detection of bacteria in cell culture production medium. *Cytometry, 71A*, 1019–1026.

Medintz, I., & Mattoussi, H. (2009). Quantum dot based resonance energy transfer ant its growing application in biology. *Physical Chemistry Chemical Physics, 11*, 17–45.

Michalet, X., Pinaud, F. F., Bentolila, L. A., Tsay, J. M., Doose, S., Li, J. J., Sundaresan, G., Wu, A. M., Gambhir, S. S., & Weiss, S. (2005). Quantum dots for live cells, in vivo imaging and diagnostics. *Science, 307*, 538–544.

Nebe-Von-Caron, G., & Muller, S. (2010). Functional single cell analyses: flow cytometry and cell sorting of microbial populations and communities. *FEMS Microbiology Reviews, 34*, 554–587.

Nebe-Von-Caron, G., Stephens, P. J., Hewitt, C. J., Powell, J. R., & Badley, R. A. (2000). Analysis of bacterial function by multi-colour fluorescence flow cytometry and single cell sorting. *Journal of Microbiological Methods, 42*, 97–114.

Shapiro, H. M. (2000). Microbial analysis at the single-cell level: Tasks and techniques. *Journal of Microbiological Methods, 42*, 3–16.

Shapiro, H. M. (2003). *Practical flow cytometry*. Hoboken: John Wiley & Sons.

Stevens, K. A., & Jaykus, L.-A. (2004). Bacterial separation and concentration from complex matrices: A review. *Critical Reviews in Microbiology, 30*, 7–24.

Stopa, P. J. (2000). The flow cytometry of bacillus anthracis spores revisited. *Cytometry, 41*, 237–244.

Su, X.-L., & Li, Y. (2004). Quantum dot biolabeling coupled with immunomagnetic separation for detection of *E. coli* O157:H7. *Analytical Chemistry, 76*, 4806–4810.

Summers, H. D., Holton, M. D., Rees, P., Williams, P. M., & Thornton, C. A. (2010). Analysis of quantum dot fluorescent stability in primary blood mononuclear cells. *Cytometry, 77A*, 933–939.

Tarnok, A. (2010). Quantum of dots. *Cytometry, 77A*, 905–906.

Venkatapathi, M., Barak, R., Ragheb, K., Banada, P. P., Lary, T., Robinson, J. P., & Hirleman, E. D. (2008). High speed classification of individual bacterial cells using a model-based light scatter system and multivariate statistics. *Applied Optics, 47*, 678–686.

Yitzhaki, S., Barnea, A., Keysary, A., & Zahavy, E. (2004). New approach for serological testing for leptospirosis by using detection of leptospira agglutination by flow cytometry light scatter analysis. *Journal of Clinical Microbiology, 42*, 1680–1685.

Yitzhaki, S., Freeman, E., Lustig, S., Keysary, A., & Zahavy, E. (2005). Double labeling and simultaneous detection of B and T cells using fluorescence nano crystal in paraffin embedded tissues. *Journal of Fluorescence, 15*, 661–665.

Zahavy, E., Fisher, M., Bromberg, A., & Olshevsky, U. (2003). Detection of FRET pair on double labeled micro-sphere and *B. anthracis* spores, by flow-cytometry. *Applied and Environmental Microbiology, 69*, 2330–2339.

Zahavy, E., Heleg-Shabtai, V., Zafrani, Y., Marciano, D., & Yitzhaki, S. (2010). Application of fluorescent nanocrystals (q-dots) as fluorescent labels for the detection of pathogenic bacteria by flow-cytometry. *Journal of Fluorescence, 20*(1), 389–399.

Zhao, Y., Ye, M., Chao, Q., Jia, N., Ge, Y., & Shen, H. (2009). Simultaneous detection of multifood-borne pathogenic bacteria based on functionalized quantum dots couples with immunomagnetic separation in food samples. *Journal of Agricultural and Food Chemistry, 57*, 517–524.

Chapter 4
Advancing Nanostructured Porous Si-Based Optical Transducers for Label Free Bacteria Detection

Naama Massad-Ivanir, Giorgi Shtenberg, and Ester Segal

Abstract Optical label-free porous Si-based biosensors for rapid bacteria detection are introduced. The biosensors are designed to directly capture the target bacteria cells onto their surface with no prior sample processing (such as cell lysis). Two types of nanostructured optical transducers based on oxidized porous Si ($PSiO_2$) Fabry-Pérot thin films are synthesized and used to construct the biosensors. In the first system, we graft specific monoclonal antibodies (immunoglobulin G's) onto a neat electrochemically-machined $PSiO_2$ surface, based on well-established silanization chemistry. The second biosensor class consists of a $PSiO_2$/hydrogel hybrid. The hydrogel, polyacrylamide, is synthesized *in situ* within the nanostructured $PSiO_2$ host and conjugated with specific monoclonal antibodies to provide the active component of the biosensor. Exposure of these modified-surfaces to the target bacteria results in "direct-cell-capture" onto the biosensor surface. These specific binding events induce predictable changes in the thin-film optical interference spectrum of the biosensor. Our studies demonstrate the applicability of these biosensors for the detection of low bacterial concentrations, in the range of 10^3–10^5 cell/ml, within minutes. The sensing performance of the two different platforms, in terms of their stability in aqueous media and sensitivity, are compared and discussed. This preliminary study suggests that biosensors based on $PSiO_2$/hydrogel hybrid outperform the neat $PSiO_2$ system.

Keywords Porous Si • Biosensor • Bacteria detection • Optical transducer • Hydrogel • Hybrid

4.1 Introduction

Porous Si (PSi) or SiO_2 matrices have emerged as an attractive and versatile material for the construction of complex functional nanostructures (Li et al. 2003, 2005; Yoon et al. 2003). PSi is typically synthesized by anodic electrochemical etching of a single-crystal Si wafer in a hydrofluoric acid (HF)

N. Massad-Ivanir
Department of Biotechnology and Food Engineering, Technion – Israel Institute of Technology, Haifa 32000, Israel

G. Shtenberg
The Interdepartmental Program of Biotechnology, Technion – Israel Institute of Technology, Haifa 32000, Israel

E. Segal (✉)
Department of Biotechnology and Food Engineering, Technion – Israel Institute of Technology, Haifa 32000, Israel

Russell Berrie Nanotechnology Institute, Technion – Israel Institute of Technology, Haifa 32000, Israel
e-mail: esegal@tx.technion.ac.il

E. Zahavy et al. (eds.), *Nano-Biotechnology for Biomedical and Diagnostic Research*,
Advances in Experimental Medicine and Biology 733, DOI 10.1007/978-94-007-2555-3_4,

based electrolyte solution. At the end of this simple and cost-effective process, a nanostructure with complex properties is obtained (Sailor and Link 2005). One of the favorable properties of this nanomaterial is its large surface area (up to 500 m^2/cm^3), which enables large amounts and a variety of biomolecular interactions, including enzymes (DeLouise et al. 2005b), DNA fragments (Zhang and Alocilja 2008) and antibodies (Bonanno and DeLouise 2007), occurring over a small working area. The resulting activated surface can be used for several biosensing applications, including the detection of DNA (Zhang and Alocilja 2008), proteins (Dancil et al. 1999; Pacholski et al. 2005, 2006; Schwartz et al. 2007), enzyme activity (DeLouise et al. 2005b) and bacteria (Alvarez et al. 2007; Chan et al. 2001; Massad-Ivanir et al. 2010; Massad-Ivanir et al. 2011; Radke and Alocilja 2005). PSi optical sensors are based on changes of photoluminescence (Chan et al. 2001; de Leon et al. 2004) or reflectivity (Archer et al. 2004; Stewart and Buriak 2000) upon exposure to the target analyte (D'Auria et al. 2006), which replace the media in the pores. A change in the refractive index of the film is observed as a modulation in the photoluminescence spectrum or as a wavelength shift in the reflectivity spectrum, respectively.

A key challenge in PSi biosensors is to effectively stabilize the nanostructure during experiments in biological solutions, as PSi oxidation and dissolution in aqueous environments lead to significant signal baseline drifts, signal loss, and ultimately to structural collapse of the PSi thin film (Jane et al. 2009; Janshoff et al. 1998; Kilian et al. 2009). Another difficulty presented by PSi transducers is the susceptibility of proteins to undergo undesired conformation changes during deposition and patterning onto the Si (Burnham et al. 2006). Thus, it was demonstrated that the use of PSi as template or as a host matrix may eliminate these issues, while providing the means for construction of complex optical structures from flexible materials, such as polymers (Bonanno and DeLouise 2009a, b; DeLouise et al. 2005a; Li et al. 2003; Massad-Ivanir et al. 2010; Segal et al. 2007; Yoon et al. 2003). Specifically, the incorporation of hydrogels offers significant advantages due to their high optical transparency, good mechanical properties, ability to store and immobilize reactive functional groups and biological compatibility (Bonanno and DeLouise 2009a, b; Massad-Ivanir et al. 2010). Recent work on (oxidized) PSi/hydrogel hybrids (Bonanno and DeLouise 2009a, b; Massad-Ivanir et al. 2010; Perelman et al. 2010; Sciacca et al. 2011; Segal et al. 2007; Wu and Sailor 2009) demonstrated the potential of these nanomaterials for application in drug delivery, sensing and biosensing.

In the present work we describe the basic considerations in designing PSi-based optical transducers for label-free detection of microorganisms e.g., bacteria. We compare two strategies for the preparation of the biosensor surface. In the first approach, neat oxidized PSi ($PSiO_2$) Fabry–Pérot thin films are used as the transducer and biofunctionalized with a monoclonal antibody (as the capture probe) using well established coupling chemistry. In the second approach, an antibody-modified $PSiO_2$/hydrogel hybrid is designed and synthesized. The hydrogel is synthesized *in situ* within the $PSiO_2$ host and conjugated with a monoclonal antibody through a biotin–streptavidin (SA) system. We demonstrate rapid detection of *E. coli* K-12 bacteria (as a model microorganism) via a "direct cell capture" approach onto these two types of PSi-based biosensors.

4.2 Materials and Methods

4.2.1 Materials

Highly-doped p-type Si wafers (0.0009 Ω-cm resistivity, <100>oriented, boron-doped) were purchased from Siltronix Corp. Aqueous HF (48%) and ethanol absolute were supplied by Merck. Acrylamide, N,N′-methylenebis(acrylamide) (BIS), bis(acryloyl)cystamine (BIS-CA), 2,2-dimethoxy-2-phenyl-acetophenone (DMAP), tris(2-carboxyethyl) phosphine (TCEP), Bis(N-succinimidyl) carbonate (SC), (3-aminopropyl)triethoxysilane (APTES), diisopropylethylamine (DIEA) and PEO-iodoacetyl biotin were obtained from Sigma Aldrich Chemicals. Streptavidin (SA), *E. coli* antibody and biotinylated *E. coli* antibody were purchased from Jackson ImmunoResearch Labs Inc. *E. coli* (K-12) was generously supplied by Prof. Sima Yaron (Technion).

4.2.2 Preparation of $PSiO_2$ Substrates

PSi Fabry-Pérot thin films are prepared by anodic etch of highly doped p-type single-crystal Si wafers in a solution of 3:1 v/v 48% aqueous hydrofluoric acid:ethanol. A constant current density of 385 mA/cm^2 is applied for 30 s. The resulting freshly-etched samples are thermally oxidized at 800°C, to create a SiO_2 matrix.

4.2.3 Preparation of $PSiO_2$/Hydrogel Hybrids

The detailed synthesis scheme of these hybrids was previously described (Massad-Ivanir et al. 2010). Briefly, an aqueous pre-gel solution contains acrylamide monomers, cross-linking agents (BIS and BIS-CA) and photoinitiator (DMAP). The pre-gel solution is cast onto the $PSiO_2$ film and allowed to infiltrate into the nanostructure. Photo-polymerization is initiated by UV irradiation (254 nm, 10 min).

4.2.4 Immobilization of Recognition Motives

4.2.4.1 Biofunctionalization of Neat $PSiO_2$ Films

Preparation of APTES-modified surfaces: A $PSiO_2$ sample is incubated with an aqueous solution of 42 mM APTES and 56 mM DIEA for 30 min. After the solution is removed, the surface is rinsed with purified water and ethanol for 10 min each and dried under a nitrogen stream.

Preparation of NHS-modified surfaces: The APTES-modified surface is immersed in a 10 mM SC solution in acetonitrile for 7 min. After the solution is removed, the surface is washed extensively with acetonitrile three times for 10 min each and dried under a nitrogen stream.

Preparation of IgG-modified surfaces: The NHS-modified surface is incubated in 100 μg/mL *E. coli* antibody solution at room temperature for 60 min. After the solution is removed, the surface is rinsed with PBS, 1 M NaCl, and PBS for 10 min each.

4.2.4.2 Biofunctionalization of $PSiO_2$/Polyacrylamide Hybrids

The hybrids are exposed to a reducing agent, TCEP (10 mM), to generate reactive thiol groups throughout the hydrogel. The activated hydrogel is then treated with PEO-iodoacetyl biotin (100 μM), a thiol-reactive biotin linker molecule, and reacted with SA (100 μg/mL). The resulting SA-modified hybrids are incubated with biotinylated *E. coli* antibody (100 μg/mL).

4.2.5 *Measurement of Interferometric Reflectance Spectra*

Interferometric reflectance spectra of the samples are collected using an Ocean Optics charge-coupled device (CCD) USB 4000 spectrometer fitted with a microscope objective lens coupled to a bifurcated fiber-optic cable. A tungsten light source is focused onto the center of the sample surface with a spot size of approximately 1–2 mm^2. Reflectivity data are recorded in the wavelength range of 400–1,000 nm, with a spectral acquisition time of 100 ms. Both illumination of the surface and detection of the reflected light are performed along an axis coincident with the surface normal. All the optical experiments are conducted in a fixed cell to ensure that the sample reflectivity is measured at the same spot during all the measurements. Spectra are collected using a CCD spectrometer and analyzed by applying fast Fourier transform (FFT).

4.2.6 *Bacteria Culture*

E. coli K-12 is cultivated in a 10 mL tube with LB medium (5 mL) (medium composition in deionized water (1 L): NaCl (5 g), yeast extract (5 g), and tryptone (10 g)). The bacteria are incubated overnight at 37°C with shaking.

4.2.7 *Bacteria Sensing*

IgG-modified $PSiO_2$, IgG-modified hybrid, neat $PSiO_2$ and unmodified hybrid (as controls) samples are incubated with *E. coli* K-12 suspensions (at concentrations ranging from 10^3 to 10^5 cell/mL) for 30 min in a fixed cell. After the bacteria suspension is removed, the cell is flushed for 30 min with a buffer solution. Optical measurements are recorded throughout the experiment. The FFT intensity changes are expressed as percentages and are calculated using the following equation:

$$\text{Intensity change } (\%) = \frac{A_1 - A_2}{A_2} \times 100 \tag{4.1}$$

where A_1 is the intensity before modification and A_2 is the intensity after modification with the bacteria suspensions.

4.3 Results and Discussion

4.3.1 *Preparation and Characterization of $PSiO_2$ Films*

The $PSiO_2$ film is prepared from a highly doped p-type single-crystal Si wafer, polished on the <100>face using an anodic electrochemical etch, at a constant current density of 385 mA/cm^2 for 30 s. The resulting freshly etched PSi template is then thermally oxidized at 800°C to create a hydrophilic $PSiO_2$ matrix. The resulting porous layer is approximately 7,880 nm thick with interconnecting cylindrical pores ranging in diameter from 60 to 100 nm and the calculated porosity is approximately 80%. The structural properties, i.e., thickness and porosity, of the $PSiO_2$ layer are thoroughly characterized by scanning electron microscopy, gravimetry (for porosity), and a spectroscopic liquid infiltration method, as previously described by Massad-Ivanir et al. (2010).

4.3.2 *Biofunctionalization of Neat $PSiO_2$ and $PSiO_2$/Hydrogel Hybrids with Antibodies*

4.3.2.1 Neat $PSiO_2$

The synthetic approach for grafting the monoclonal antibodies (IgG) onto neat $PSiO_2$ surfaces is based on a well established silanization technology (Xia et al. 2006), as previously described (Massad-Ivanir et al. 2011).

4.3.2.2 $PSiO_2$/Hydrogel Hybrids

Biofunctionalization of the hybrids is achieved by exposure of the $PSiO_2$/di-sulfide linked hydrogel to a reducing (TCEP) to generate reactive thiol groups throughout the hydrogel phase. Next, the activated hydrogel is incubated with a SH-reactive biotin linker molecule, followed by attachment of SA to the hydrogel via the SH-linked biotin. Functionalization is accomplished through the bioconjugation of biotinylated *E. coli* monoclonal antibodies to the SA linker molecule.

4.3.3 *Stability of $PSiO_2$–Based Transducers in Aqueous Media*

The optical readout stability of the two different transducers (neat $PSiO_2$ and hybrid) under flow conditions is characterized by long buffer (PBS) flow experiments of several hours. The samples are placed in a plexiglas custom-made flow cell, in which a buffer solution is delivered at a constant buffer flow of 0.5 mL/min for 5 h and the reflectivity spectra is collected every 15 s.

The reflectivity spectrum of the thin PSi layer consists of a series of interference fringes that result from a Fabry–Pérot interference. This fringe pattern arises from reflections at the top and at the bottom of the film, so that the measurement is made over the entire volume of the system. The maxima of these fringes are governed by the following relationship (Fabry–Pérot equation) (Sailor and Link 2005):

$$m\lambda = 2nL \tag{4.2}$$

The effective optical thickness (EOT) of the sample refers to the 2 nL term in the Fabry–Pérot formula (where m is an integer, n is the average refractive index, L is the thickness of the film, and λ is the wavelength of the incident light). A change in the average refractive index (n) leads to a shift in the observed reflectivity spectrum that correlates with EOT changes.

Figure 4.1 shows the results of the buffer flow experiments for both silanized-$PSiO_2$ and hybrid transducers. Changes in the EOT values and the reflected light intensity with time are depicted in Fig. 4.1a and b, respectively. This study demonstrates that the optical readout of the hybrid transducer is highly stable throughout the entire experiment. Changes of less than 0.5% in the intensity and 10 nm in the EOT values are recorded. Thus, a robust optical readout with no evident baseline drift is obtained. On the contrary, the silanized-$PSiO_2$ transducer exhibits significant baseline drifts of both

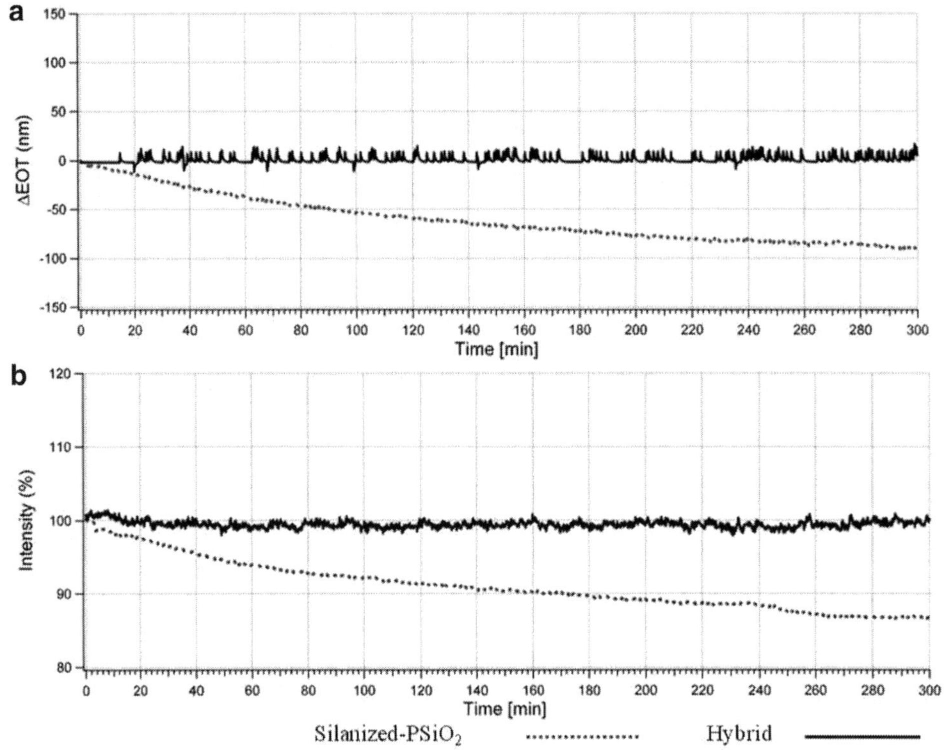

Fig. 4.1 Buffer flow experiments onto silanized-$PSiO_2$ and $PSiO_2$/polyacrylamide hydrogel hybrid transducers. (**a**) EOT changes with time and (**b**) Intensity changes with time. The reflectivity spectra of the samples is collected every 15 s under constant PBS buffer flow of 0.5 mL/min for 5 h

the EOT and the intensity signals with time. The EOT value of the film decrease by approximately 90 nm within 5 h and the reflected intensity drops by 13%. These significant changes in the optical readout are ascribed to time-dependent degradation of the porous scaffold (Janshoff et al. 1998; Massad-Ivanir et al. 2010; Pacholski et al. 2006). Thus, these experiments clearly show that the presence of the hydrogel within and delicate and highly porous SiO_2 scaffold stabilize the nanostructure under aqueous conditions. Similar results were recently reported by Cunin et al. (Sciacca et al. 2011) for a PSi/chitosan hybrid used as an optical transducer for the detection of carboxylic acid-containing drugs in water.

4.3.4 Optical Detection of *E. coli* Bacteria

Our biosensing approach is based on monitoring changes in the light reflected from the biofunctionalized $PSiO_2$ or hybrid nanostructures. Changes in the amplitude (intensity) of the FFT peak of the biosensor are correlated to specific immobilization of bacteria cells, "direct cell capture" onto the transducer surface via antibody-antigen interactions. Since *E. coli* bacteria are large biological species, with typical dimensions of 0.8–2 μm (Sundararaj et al. 2004), binding of bacteria will occur only on the biosensor surface and not inside the porous nanostructure. Moreover, we expect that their presence on the surface will scatter the light, resulting in a significant decrease in the reflected light intensity (Alvarez et al. 2007; Massad-Ivanir et al. 2010, 2011). An experimental

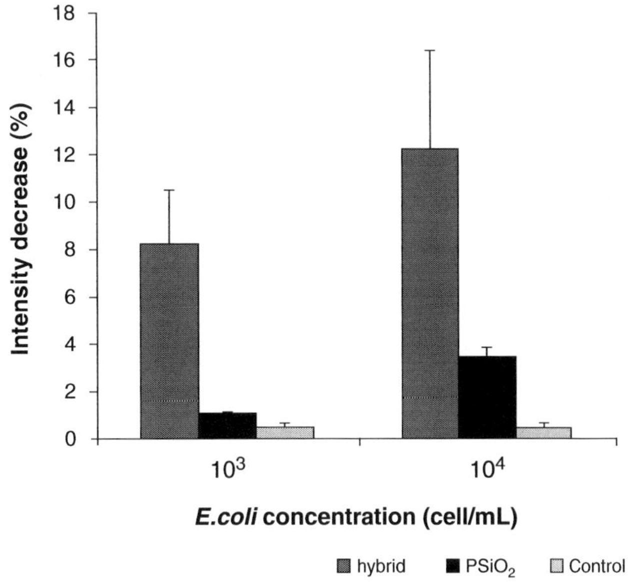

Fig. 4.2 *E. coli* K-12 biosensing experiments. IgG-modified $PSiO_2$ and hybrid are incubated with different concentrations of *E. coli* suspensions and the maximal intensity decrease value is presented. Control experiments, incubation of unmodified $PSiO_2$ and hybrid surfaces with bacteria suspensions

setup for continuous monitoring of the reflectivity spectrum of the hybrids upon incubation with bacteria suspensions is designed and constructed. The biosensors are exposed to *E. coli* K-12 suspensions with different concentrations, ranging from 10^3 to 10^5 cell/mL. The incubation time was set to 30 min, after which the samples were washed with a buffer solution for 30 min. Figure 4.2 displays the optical response of the different biosensors upon introduction to different concentrations of *E. coli* suspensions. Indeed, both biosensors exhibit a decrease in intensity upon exposure to *E. coli*, while insignificant changes in the FFT spectrum are recorded for the unmodified surfaces. In order to confirm that this intensity change results from bacteria capture onto IgG-modified surfaces, the biosensors are studied under a light microscope immediately after the biosensing experiment. Indeed, immobilized bacteria cells are observed onto the biosensor surface, while no cells are observed onto the unmodified surfaces (data not shown). Moreover, the intensity signals of the two biosensors are proportional to the bacteria suspension concentration (see Fig. 4.2). As expected, exposure of the biosensors to a lower bacteria concentration (10^3 cells/mL) results in a signal decrease; i.e., a smaller change in intensity is observed. When comparing the response of the two biosensors ($PSiO_2$ vs. hybrid) to *E. coli* exposure, the hybrid-based biosensors exhibit significantly higher intensity changes (~four fold higher). These results indicate that the hybrid system outperforms the $PSiO_2$ biosensors in terms of their sensitivity. It is well established that the immobilization step is crucial in maintaining the conformation and the immunoactivity of the IgG molecules (Somasundaran 2006). We assume that the presence of the hydrogel (a soft organic component) maintains the proteins desired conformation during the immobilization and patterning onto the biosensor surface.

In terms of the sensitivity of the different biosensors, these preliminary experiments show relatively low detection limits of 10^3 and 10^4 cell/mL for the modified hybrids and $PSiO_2$ biosensing platforms, respectively. For comparison, the detection limit of current state-of-the-art surface plasmon resonance (SPR) biosensors is in the range of 10^2–10^6 cells/mL (Dudak and Boyaci 2007, 2009; Skottrup et al. 2008; Taylor et al. 2008). Moreover, the response time of these biosensors to bacteria exposure is comparable to that of SPR techniques.

4.4 Conclusions

Two classes of label-free optical biosensors for bacteria detection are synthesized and characterized. The first platform is based on a $PSiO_2$ nanostructure (Fabry–Pérot thin film) and the second consists of a $PSiO_2$/polyacrylamide hydrogel hybrid. The different nanostructures' surfaces are biofunctionalized with IgG, as a capture probe, and their potential applicability as a biosensor for bacteria detection is demonstrated. The hybrids show improved optical readout stability under aqueous conditions. Moreover, the presence of the hydrogel also enables desired conformation of the antibodies onto the biosensor surface during deposition and patterning, resulting in a higher sensitivity. This proof-of-concept work demonstrates a simple and sensitive detection scheme of bacteria via a "direct cell capture" approach. Our preliminary biosensing experiments demonstrate a detection limit of 10^3–10^4 cells/mL for *E. coli* and a response time of several minutes. We are currently exploring several approaches to enhance the sensitivity of these biosensors, including improving optimization of the antibody concentration and orientation, enhancement of the coupling chemistry and use antibody fragments.

Acknowledgements This work was supported by Marie Curie European Reintegration Grant, The Israel Science Foundation (grant No. 1118/08). E.S gratefully acknowledges the generous financial support of the Technion and the Russell Berrie Nanotechnology Institute.

References

Alvarez, S. D., Schwartz, M. P., Migliori, B., Rang, C. U., Chao, L., & Sailor, M. J. (2007). Using a porous silicon photonic crystal for bacterial cell-based biosensing. *Physica Status Solidi a–Applications and Materials Science, 204*, 1439–1443.

Archer, M., Christophersen, M., Fauchet, P. M., Persaud, D., & Hirschman, K. D. (2004). Electrical porous silicon microarray for DNA hybridization detection. *Micro- and Nanosystems, 782*, 385–391.

Bonanno, L. M., & Delouise, L. A. (2007). Steric crowding effects on target detection in an affinity biosensor. *Langmuir, 23*, 5817–5823.

Bonanno, L. M., & Delouise, L. A. (2009a) Design of a hybrid amine functionalized polyacrylamide hydrogel-porous silicon optical sensor. *Proceedings of SPIE, 7167*, 71670F1-11.

Bonanno, L. M., & Delouise, L. A. (2009b). Optical detection of polyacrylamide swelling behavior in a porous silicon sensor. *Materials Research Society Symposium. Proceeding., 1133*, 1133-AA01-05.

Burnham, M. R., Turner, J. N., Szarowski, D., & Martin, D. L. (2006). Biological functionalization and surface micropatterning of polyacrylamide hydrogels. *Biomaterials, 27*, 5883–5891.

Chan, S., Horner, S. R., Fauchet, P. M., & Miller, B. L. (2001). Identification of gram negative bacteria using nanoscale silicon microcavities. *Journal of the American Chemical Society, 123*(47), 11797–11798.

Dancil, K.-P. S., Greiner, D. P., & Sailor, M. J. (1999). A porous silicon optical biosensor: Detection of reversible binding of IgG to a protein A-modified surface. *Journal of the American Chemical Society, 121*, 7925–7930.

D'Auria, S., de Champdore, M., Aurilia, V., Parracino, A., Staiano, M., Vitale, A., Rossi, M., Rea, I., Rotiroti, L., Rossi, A. M., Borini, S., Rendina, I., & de Stefano, L. (2006). Nanostructured silicon-based biosensors for the selective identification of analytes of social interest. *Journal of Physics. Condensed Matter, 18*, S2019–S2028.

de Leon, S. B., Sa'Ar, A., Oren, R., Spira, M. E., & Yitzchaik, S. (2004). Neurons culturing and biophotonic sensing using porous silicon. *Applied Physics Letters, 84*, 4361–4363.

Delouise, L. A., Fauchet, P. M., Miller, B. L., & Pentland, A. A. (2005a). Hydrogel-supported optical-microcavity sensors. *Advanced Materials, 17*, 2199–2203.

Delouise, L. A., Kou, P. M., & Miller, B. L. (2005b). Cross-correlation of optical microcavity biosensor response with immobilized enzyme activity. Insights into biosensor sensitivity. *Analytical Chemistry, 77*, 3222–3230.

Dudak, F. C., & Boyaci, I. H. (2007). Development of an immunosensor based on surface plasmon resonance for enumeration of *Escherichia coli* in water samples. *Food Research International, 40*, 803–807.

Dudak, F. C., & Boyaci, I. H. (2009). Rapid and label-free bacteria detection by surface plasmon resonance (SPR) biosensors. *Biotechnology Journal, 4*, 1003–1011.

Jane, A., Dronov, R., Hodges, A., & Voelcker, N. H. (2009). Porous silicon biosensors on the advance. *Trends in Biotechnology, 27*, 230–239.

Janshoff, A., Dancil, K. P. S., Steinem, C., Greiner, D. P., LIN, V. S. Y., Gurtner, C., Motesharei, K., Sailor, M. J., & Ghadiri, M. R. (1998). Macroporous p-type silicon Fabry-Perot layers. Fabrication, characterization, and applications in biosensing. *Journal of the American Chemical Society, 120*, 12108–12116.

Kilian, K. A., Boecking, T., & Gooding, J. J. (2009). The importance of surface chemistry in mesoporous materials: Lessons from porous silicon biosensors. *Chemical Communications*, 630–640.

Li, Y. Y., Cunin, F., Link, J. R., Gao, T., Betts, R. E., Reiver, S. H., Chin, V., Bhatia, S. N., & Sailor, M. J. (2003). Polymer replicas of photonic porous silicon for sensing and drug delivery applications. *Science, 299*, 2045–2047.

Li, Y. Y., Kollengode, V. S., & Sailor, M. J. (2005). Porous silicon/polymer nanocomposite photonic crystals by microdroplet patterning. *Advanced Materials, 17*, 1249–1251.

Massad-Ivanir, N., Shtenberg, G., Zeidman, T., & Segal, E. (2010). Construction and characterization of porous SiO_2/hydrogel hybrids as optical biosensors for rapid detection of bacteria. *Advanced Functional Materials, 20*, 2269–2277.

Massad-Ivanir, N., Shtenberg, G., Tzur, A., Krepker, A. M., & Segal, E. (2011). Engineering nanostructured porous SiO_2 surfaces for bacteria detection via "direct cell capture". *Analytical Chemistry, 83*, 3282–3289.

Pacholski, C., Sartor, M., Sailor, M. J., Cunin, F., & Miskelly, G. M. (2005). Biosensing using porous silicon double-layer interferometers: Reflective interferometric Fourier transform spectroscopy. *Journal of the American Chemical Society, 127*, 11636–11645.

Pacholski, C., Yu, C., Miskelly, G. M., Godin, D., & Sailor, M. J. (2006). Reflective interferometric Fourier transform spectroscopy: A self-compensating label-free immunosensor using double-layers of porous SiO_2. *Journal of the American Chemical Society, 128*, 4250–4252.

Perelman, L. A., Moore, T., Singelyn, J., Sailor, M. J., & Segal, E. (2010). Preparation and characterization of a pH- and thermally responsive poly(N-isopropylacrylamide-co-acrylic acid)/porous SiO_2 hybrid. *Advanced Functional Materials, 20*, 826–833.

Radke, S. M., & Alocilja, E. C. (2005). A microfabricated biosensor for detecting foodborne bioterrorism agents. *IEEE Sensors Journal, 5*, 744–750.

Sailor, M. J., & Link, J. R. (2005). "Smart dust": Nanostructured devices in a grain of sand. *Chemical Communications*, 1375–1383.

Schwartz, M. P., Alvarez, S. D., & Sailor, M. J. (2007). Porous SiO_2 interferometric biosensor for quantitative determination of protein interactions: Binding of protein a to immunoglobulins derived from different species. *Analytical Chemistry, 79*, 327–334.

Sciacca, B., Secret, E., Pace, S., Gonzalez, P., Geobaldo, F., Quignard, F., & Cunin, F. (2011). Chitosan-functionalized porous silicon optical transducer for the detection of carboxylic acid-containing drugs in water. *Journal of Materials Chemistry, 21*, 2294–2302.

Segal, E., Perelman, L. A., Cunin, F., di Renzo, F., Devoisselle, J. M., Li, Y. Y., & Sailor, M. J. (2007). Confinement of thermoresponsive hydrogels in nanostructured porous silicon dioxide templates. *Advanced Functional Materials, 17*, 1153–1162.

Skottrup, P. D., Nicolaisen, M., & Justesen, A. F. (2008). Towards on-site pathogen detection using antibody-based sensors. *Biosensors and Bioelectronics, 24*, 339–348.

Somasundaran, P. (2006). *Encyclopedia of surface and colloid science*. Boca Raton: CRC Press.

Stewart, M. P., & Buriak, J. M. (2000). Chemical and biological applications of porous silicon technology. *Advanced Materials, 12*, 859–869.

Sundararaj, S., Guo, A., Habibi-Nazhad, B., Rouani, M., Stothard, P., Ellison, M., & Wishart, D. S. (2004). The CyberCell Database (CCDB): A comprehensive, self-updating, relational database to coordinate and facilitate in silico modeling of Escherichia coli. *Nucleic Acids Research, 32*, D293–D295.

Taylor, A. D., Ladd, J., Homola, J., & Jiang, S. (2008). Surface plasmon resonance (SPR) sensors for the detection of bacterial pathogens. In Z. Mohammed, E. Souna, & T. Anthony (Eds.), *Principles of bacterial detection: Biosensors, recognition receptors and microsystems*. New York: Springer.

Wu, J., & Sailor, M. (2009). Chitosan hydrogel-capped porous SiO_2 as a pH responsive nano-valve for triggered release of insulin. *Advanced Functional Materials, 19*, 733–741.

Xia, B., Xiao, S. J., Guo, D. J., Wang, J., Chao, M., Liu, H. B., Pei, J., Chen, Y. Q., Tang, Y. C., & Liu, J. N. (2006). Biofunctionalisation of porous silicon (PS) surfaces by using homobifunctional cross-linkers. *Journal of Materials Chemistry, 16*, 570–578.

Yoon, M. S., Ahn, K. H., Cheung, R. W., Sohn, H., Link, J. R., Cunin, F., & Sailor, M. J. (2003). Covalent crosslinking of 1-D photonic crystals of microporous Si by hydrosilylation and ring-opening metathesis polymerization. *Chemical Communications*, 680–681.

Zhang, D., & Alocilja, E. C. (2008). Characterization of nanoporous silicon-based DNA biosensor for the detection of salmonella enteritidis. *IEEE Sensors Journal, 8*, 775–780.

Chapter 5
Gold Fibers as a Platform for Biosensing

Sharon Marx, Moncy V. Jose, Jill. D. Andersen, and Alan J. Russell

Abstract A new form of high surface bioelectrode based on electrospun gold microfiber with immobilized glucose oxidase was developed. The gold fibers were prepared by electroless deposition of gold nanoparticles on a poly(acrylonitrile)-$HAuCl_4$ electrospun fiber. The material was characterized using electron microscopy, XRD and BET, as well as cyclic voltammetry and biochemical assay of the immobilized enzyme. The surface area of the gold microfibers was 2.5 m^2/g. Glucose oxidase was covalently crosslinked to the gold surface using cystamine monolayer and glutardialdehyde, and portrayed characteristic catalytic currents for oxidizing glucose using a ferrocene methanol mediator. Limit of detection of glucose is 0.1 mM. The K_m of the immobilized enzyme is 10 mM, in accordance with other reports of immobilized glucose oxidase. The microfiber electrode was reproducible and showed correlation between fiber weight, cathodic current and enzymatic loading.

Keywords Biosensor • Electrode • Electrospinning • Enzyme immobilization • Glucose oxidase • Gold fibers

5.1 Introduction

Biosensor platforms are the basic interface between the biorecognition sensing element – the immobilized enzyme, antibody or receptor, and the transduction system. Electrochemical biosensors rely mostly on the immobilization of redox enzymes on the surface of conductive surfaces such as gold and carbon. The immobilization of the enzyme on the surface is a crucial step in the design of the sensor, and it is important to make sure that the immobilization will render the enzyme electroactive, stable and with high population density. A limit to the packing of enzyme on the electrode surface is set by the micro structured surface area of the electrode. Methods for increasing the surface area of gold electrodes have been used in the past (Gilardi and Fantuzzi 2001; Shoham et al. 1995; Wang et al. 2008) and showed that chemical etching or attachment of gold nanoparticles increase the available

S. Marx (✉)
Department of Physical Chemistry, Israel Institute for Biological Research, Ness Ziona 74100, Israel
e-mail: sharonma@iibr.gov.il

M.V. Jose • J.D. Andersen • A.J. Russell
McGowan Institute for Regenerative Medicine, University of Pittsburgh, Pittsburgh, PA 15219, USA

E. Zahavy et al. (eds.), *Nano-Biotechnology for Biomedical and Diagnostic Research*,
Advances in Experimental Medicine and Biology 733, DOI 10.1007/978-94-007-2555-3_5,

surface area for enzyme attachment and thus the resulting enzyme loading per area unit increases as well. Other examples of increasing the surface area of gold electrodes are using three-dimensional structures that are built in the "bottoms-up" approach, and include the intricate architecture of assembling gold nano and microparticles for the formation of conducting surfaces. A promising approach is the use of gold microfibers. The high aspect ratio of these fibers ensures high surface area of the resulting material. The fabrication of microfibers of gold can be achieved through techniques like patterning, etching and extrusion, and through electrospinning.

Electrospinning has been used recently to produce new fibrous materials for a wide variety of applications- from filtering (Huang et al. 2003) to tissue scaffolding (Xu et al. 2004; Li et al. 2002). Basically, fibers are spun by applying high voltage (typically over 5 KV) on a drop of s polymer solution. The polymer drop is charged, and due to electrostatic repulsion from the source, the drop is elongated towards a grounded target. If the viscosity of the polymer solution is high enough the drop stretches via a whipping process until it hits the grounded collecting plate. The solvent evaporates in the air during the stretching process, thus resulting in a nano to micrometer scale fiber diameter (based on solution viscosity, applied voltage and other parameters). Continuous gold fibers have been produced by electrospinning, and have been characterized to be polycrystalline and conductive (Guo et al. 2008; Pol et al. 2008). In this paper, we go a step further from the mere fabrication of a gold microfiber, and use this fibrous metal as an electrode on which an enzyme is immobilized and is used as a biosensor. For the demonstration of the performance of the gold fibers as a platform of enzyme immobilization and biosensing, we chose the well-studied enzyme glucose oxidase, due to its well defined electrochemistry and facile immobilization (Wilson and Turner 1992).

5.2 Experimental

5.2.1 Materials

Poly(acrylonitrile) mw 150,000 (PAN), $HAuCl_4$ trihydrate, Ferrocene methanol, glucose, hydroxyl amine hydrochloride, $NaBH_4$, NaI, glutardialdehyde (25% in H_2O), Glucose Oxidase from *aspergillus niger* type X, 136 u/mg, DMF, were form sigma-Aldrich; cystamine and Gold wire (1 mm diameter) were purchased from Acros.

5.2.2 Electrospinning of PAN/$HAuCl_4$ Composite

A solution containing PAN 8% (w/w) and $HAuCl_4$ 10% w/w was prepared by dissolving the solids in DMF and heating at 80°C. Electrospinning was performed using an in-house built instrument consisting of a power sources (Gamma High voltage, Fl), and syringe pump (World Precision Instruments, Fl). Electrospinning voltage was 20 KV, polymer flow rate was 1 ml/h, and the tip-to collector distance was 8 cm. the fibers were collected on Al foil or on a gold rod rotated by a motor (in-house built set up) at 60 rpm.

5.2.3 Fabrication of Gold Microfibers

The resulting fibers were first reduced by 1 mM $NaBH_4$, washed with dilute HCl and then H_2O. Gold was electroless deposited on the fiber by incubating the fiber for 10–30 min in an aqueous solution containing 0.02 mg/ml hydroxylamine hydrochloride and 0.1 mg/ml $HAuCl_4$. This was repeated five times.

5.2.4 *Enzyme Modification*

The fibers were incubated in 0.1 M cystamine for 2 h, washed with H_2O and the incubated for 1 h in 5% glutardialdehyde. The fibers were then incubated in Gox, (5 mg/ml in phosphate buffer pH 6.5, 50 mM) overnight at 4°C. Gox assay on the fibers was measured using the O-dianisidine assay method.

5.2.5 *Electrochemical Characterization*

Electrochemical experiments were run on a CHI 660 electrochemical station, equipped with a standard 3-electrode cell. Auxiliary electrode was a Pt wire, and Ag/AgCl was used as a reference electrode. Cyclic voltammetry to measure the response to glucose was run by scanning between 0 and 0.4 V vs. Ag/AgCl at a scan rate of 5 mV/s, after adding 2.5×10^{-4} M ferrocene methanol as a diffusional electron mediator.

5.2.6 *Fiber Electrochemical Surface Area Measurement*

The electrochemical surface area of the gold microfibers was determined by using I^--IO_3^-adsorption oxidation method.

5.2.7 *Fiber Physical Characterization*

Scanning Electron Microscope (SEM) images were performed on carbon coated fibers, at 10 KV acceleration voltage using a JEOL 6330F SEM. Fiber diameter and gold particle size distribution were measured using ImageJ software (freeware, downloaded from rsbweb.nih.gov/ij/) on SEM images, using at least 50 objects for reliable statistics.

5.3 Results and Discussion

5.3.1 *Material Characterization*

Gold microfibers have been produced and partially characterized by other groups. In this report we expand the physical characterization of the electrospun-electroless deposition gold fibers, and include detailed surface area measurements and detailed microscopical observations. The electrospun fibers were imaged using SEM using both secondary emission and backscatter detectors to image both the topography and the difference in material density of the resulting electrospun fiber. The as-spun fibers contained defined domains of $HAuCl_4$ (Fig. 5.1a). These domains are created during the evaporation of the solvent which result in partial crystallization of the gold chloride salt and phase separation. The gold chloride is reduced initially by $NaBH_4$, a process that leaves small metallic gold regions on the fibers. These small metallic clusters serve as the nucleation seeds for the electroless gold deposition process that follows. During the electroless deposition process, Au^{3+} is reduced by the mild reducer

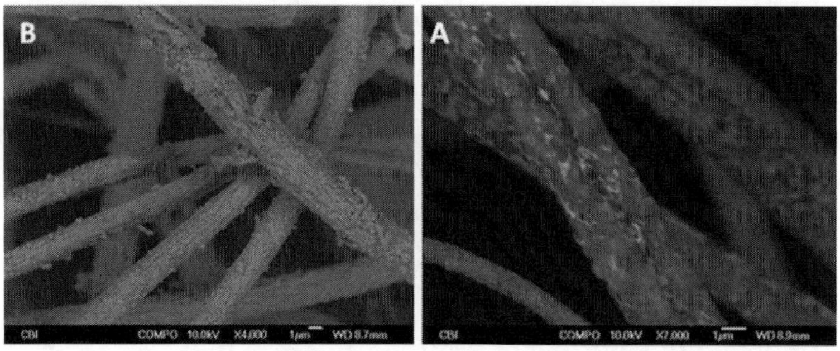

Fig. 5.1 (**a**) SEM of as-spun PAN-$HAuCl_4$ polymer; (**b**) SEM of the same polymer after electroless deposition of gold

hydroxylamine to metallic gold (Guo et al. 2008; Liang et al. 2003; Okinaka and Hoshino 1995). In the presence of a nucleation seed, i.e. the gold particle in the fiber, the gold is deposited on the seed and the gold microcrystal grows Fig. 5.1b. From the SEM images, it seems that the gold microcrystals are ordered, and form linear crystalline arrays. The proximity of the gold microcrystals indicates that the material is metallic. The XRD analysis (not shown) shows four characteristic diffraction peaks corresponding to the (111), (200), (220) and (311) planes of a face-centered cubic gold crystal. After the electroless deposition procedure, the mean diameter of the fibers is 0.99 ± 0.49 μm. The average size of the gold particles encrusting the polymer fiber is 93.1 ± 27.7 nm. The remaining gold particles in the core of the fiber that were not exposed to the electroless deposition process remain small and have an average size of 22.3 ± 6.6 nm.

The electrochemical surface area was measured by adsorbing a monolayer of iodide ions on the gold surface and the measurement of their oxidation to Iodate (IO_3^-) species. The resulting electrochemical surface area was 2.5 m^2/g. This indicates that the surface of the fiber is accessible to diffusional species – such as glucose or other dissolved analytes, and also that immobilization of enzymes on the surface of the fiber will make use of the entire surface area of the fiber and not just the area that is exposed to the solution.

5.3.2 Enzyme Immobilization on the Fiber

Glucose oxidase (Gox) was immobilized on the gold fibers using conventional crosslinking chemistry using a self-assembled monolayer of cystamine as the chemical anchor to gold. Gox was crosslinked to the amine monolayer using glutardialdehyde. The enzymatic activity of the immobilized Gox was assayed using o-dianisidine assay directly on the enzyme-fiber assembly. The coverage of the enzyme electrode was found to be ca. 0.005 units/cm^2, which corresponds to ca. 8.6% of surface coverage (based on the Gox footprint of 80 nm^2 (Hecht et al. 1993)).

5.4 Electrochemical Evaluation of Gox on Gold Fibers

Gox electrochemical activity was measured using the catalytic cathodic current observed when the electron mediator is regenerated by the FAD active site after glucose oxidation. The electron mediator used in this study was ferrocene methanol, which is commonly used as a mediator for Gox (Yokoyama and Kayanima 1998). Figure 5.2 shows a typical response of the Gox-gold fiber electrode to increasing

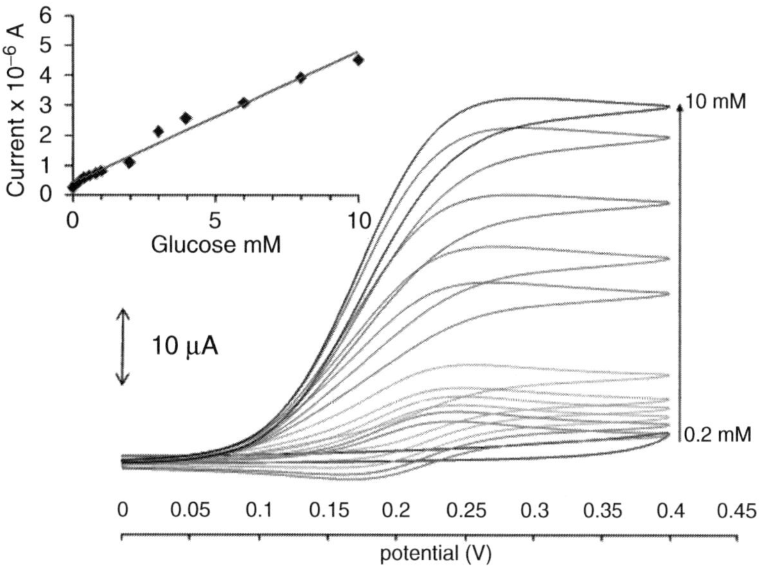

Fig. 5.2 Cyclic voltammetry of Gox immobilized on gold fiber electrode, in 0.05 M phosphate buffer pH = 7.0, containing 2.5 × 10^{-4} M ferrocene methanol. Scan rate = 5 mV/s. *Inset*: The calibration graph of the response of the electrode to increasing glucose concentrations, measured for the peak of the cathodic peak

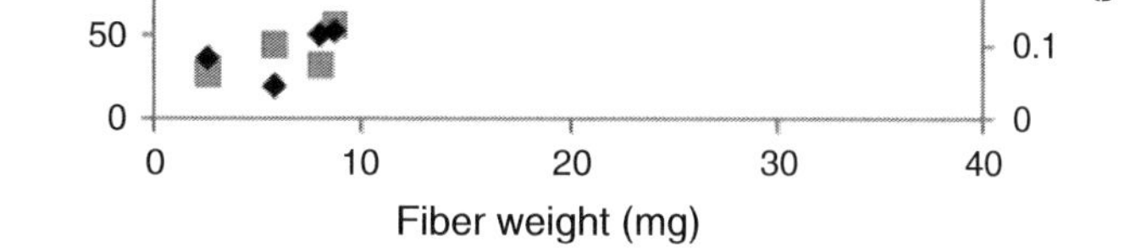

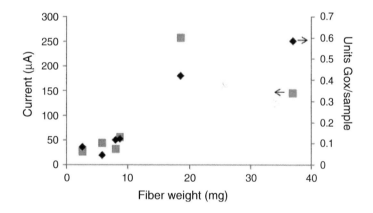

Fig. 5.3 Correlation between the weight of the gold fiber electrode and the cathodic current (■) and the Gox loading on each electrode (♦). Measured in 5 mM glucose, 5 mV/s scan rate, solution containing 2.5 × 10^{-4} M ferrocene methanol

glucose concentrations. Linear response to glucose concentration in the range of 0.1–20 mM is observed, with limit o detection of 0.1 mM. The Michaelis-Menten constant, K_m, of the immobilized Gox on the fiber electrode was calculated from the electrode responses (Fig. 5.2, inset), and was found to be 10 mM. This value is in good agreement with K_m of immobilized Gox in similar systems (Hale et al. 1991), and is ca. 10 times larger than the K_m of Gox measured in solution (typically, 1 mM). The reason for the increased K_m is the slower diffusion of the substrate, glucose, to the immobilized enzyme relative to the uninhibited diffusion in a homogeneous system.

The stability of the enzyme electrode was measured by repeatedly measuring the current at 10 mM glucose concentration, and it was found that the variability in the measurements was 13%, with an average current of 30.3 ± 0.4 µA. Figure 5.3 shows the relation between the enzyme loading as measured by Gox assay and the catalytic current measured by glucose oxidation mediated by ferrocene methanol.

Enzyme immobilization linearly increased as the weight of the fiber increased, and hence the available surface area increased as well. This was also satisfactorily correlated to the increase in the cathodic current in the electro-oxidation of glucose via ferrocene methanol.

5.5 Summary and Conclusions

A platform for biosensing based on gold microfiber has been developed. The material was characterized using a myriad of physical methods, and was shown to portray high surface area, suitable for biosensing and other electrochemical applications. The high surface area is beneficial to all application that requires high surface area and rapid diffusion of substrates in gas or liquid phase. The gold surface is one of the more versatile materials that are used in the biosensor field. There are numerous ways to immobilized enzymes to its surface using a variety of cross linking methodologies. In this paper, we show the covalent immobilization of glucose oxidase to the surface of the electrode and the application of this new electrode as a basic glucose sensor.

References

Gilardi, G., & Fantuzzi, A. (2001). Manipulating redox systems: Application to nanotechnology. *Trends in Biotechnology, 19*, 468–476.

Guo, B., Zhao, S., Han, G., & Zhang, L. (2008). Continuous thin gold films electroless deposited on fibrous mats of polyacrylonitrile and their electrocatalytic activity towards the oxidation of methanol. *Electrochimica Acta, 53*, 5174–5179.

Hale, P. D., Boguslavsky, L. I., Inagaki, T., Karan, H. I., Lee, H. S., Skotheim, T. A., & Okamoto, Y. (1991). Amperometric glucose biosensors based on redox polymer mediated electron transfer. *Analytical Chemistry, 63*, 677–682.

Hecht, H. J., Schomburg, D., Kalisz, H., & Schmid, R. D. (1993). The 3D structure of glucose oxidase from aspergillus Niger. Implications for the use of GOD as a biosensor enzyme. *Biosensors & Bioelectronics, 8*, 197–203.

Huang, Z.-M., Zhang, Y.-Z., Kotaki, M., & Ramakrishna, S. (2003). A review on polymer nanofibers by electrospinning and their applications in nanocomposites. *Composite Science Technology, 63*, 2223–2253.

Li, W.-J., Laurencin, C. T., Caterson, E. J., Tuan, R. S., & Ko, F. K. (2002). Electrospun nanofibrous structure: A novel scaffold for tissue engineering. *Journal of Biomedical Materials Research, 60*, 613–621.

Liang, Z., Susha, A., & Caruso, F. (2003). Gold nanoparticle based core-shell and hollow spheres and ordered assemblies thereof. *Chemistry of Materials, 15*, 3176–3183.

Okinaka, Y., & Hoshino, M. (1995). Some recent topics in gold plating for electronics applications. *Gold Bulletin, 31*, 3.

Pol, V. G., Koren, E., & Zaban, A. (2008). Fabrication of continuous conducting gold wires by electrospinning. *Chemistry of Materials, 20*, 3055–3062.

Shoham, B., Migron, Y., Riklin, A., Willner, I., & Tartakovsky, B. (1995). A bilirubon biosensor based on a multilayer network enzyme electrode. *Biosensors & Bioelectronics, 10*, 341–352.

Wang, L., Wang, L., Di, J., & Tu, Y. (2008). Disposable biosensor based on immobilization of glucose oxidase at gold nanoparticles electrodeposited on ITO electrode. *Sensors and Actuators B, 135*, 283–288.

Wilson, R., & Turner, A. P. F. (1992). Glucose oxidase - the ideal enzyme. *Biosensors & Bioelectronics, 7*, 165–185.

Xu, C. Y., Inai, R., Kotaki, M., & Ramakrishna, S. (2004). Aligned biosegradable nanofibrouse structure: A potential scaffold for blood vessel engineering. *Biomaterials, 25*, 877–886.

Yokoyama, K., & Kayanima, Y. (1998). Cyclic voltammetry simulation for electrochemically mediated enzyme reaction and determination of enzyme kinetic constant. *Analytical Chemistry, 70*, 3368–3376.

Chapter 6
Surface-Enhanced Raman Spectroscopy of Organic Molecules Adsorbed on Metallic Nanoparticles

Vered Heleg-Shabtai, Adi Zifman, and Shai Kendler

Abstract The improvements in Raman instrumentation have led to the development of portable, simple to operate, Raman instruments that can be used for on-site analysis of substances relevant for homeland security purposes such as chemical and biological warfare and explosives materials.

Raman spectroscopy, however, suffers from limited sensitivity which can be overcome by Surface-Enhanced Raman Spectroscopy (SERS). SERS can enhance the Raman signal of a target molecule by 6–10 orders of magnitude. The increased sensitivity, together with Raman's molecular recognition capabilities and the availability of portable Raman instruments make SERS a powerful analytical tool for on site detection.

In this work we studied the effect of target molecules and SERS-active substrate properties on the obtained SERS, using a field portable Raman spectrometer. Also reported herein is the SERS detection of the chemical warfare agent sulfur mustard (HD, 2,2 dichloroethyl sulfide). This study may serve as a basis for the development of SERS platform for homeland security purposes.

Keywords SERS • Raman • CWA • HD • Nanoparticles

6.1 Introduction

Raman spectroscopy provides valuable information regarding the vibration spectrum of a wide range of materials. Using an appropriate data base and algorithm, this information can be used to identify the content of a sample. The advances in Raman instrumentation have led to the development of portable, easy to use Raman instruments that can be used for on-site analysis of substances that are relevant for homeland security purposes such as chemical and biological warfare and explosives materials (Moore and Scharff 2009).

Utilization of this technique, however, is limited due to two main drawbacks: low sensitivity of the Raman Effect and the presence of background fluorescence. These limitations can be overcome by the enhancement method of surface-enhanced Raman spectroscopy (SERS) (Smith 2008; Dijkstra et al. 2005).

SERS is a surface effect that is observed when molecules are found in the vicinity of certain metal substrates and is arises mainly as a result of enhanced localized electric field generated near roughened

V. Heleg-Shabtai (✉) • A. Zifman • S. Kendler
Department of Physical Chemistry, Israel Institute for Biological Research, P.O. Box 19, Ness-Ziona 74100, Israel
e-mail: veredhs@iibr.gov.il

E. Zahavy et al. (eds.), *Nano-Biotechnology for Biomedical and Diagnostic Research*, Advances in Experimental Medicine and Biology 733, DOI 10.1007/978-94-007-2555-3_6,

metallic surfaces. This mechanism is referred to as electromagnetic mechanism (EMM) (Moskovits 1985; Brolo et al. 1997) and is due to the coupling of the incident radiation to the surface plasmons. EMM together with other mechanisms such as chemical mechanism (Otto 2005), increase the Raman signal of the target molecule by 6–10 orders of magnitude. The increased sensitivity, together with Raman's molecular recognition capabilities and the availability of portable Raman instruments make SERS a powerful analytical tool for field detection.

Different types of SERS-active substrates have been developed over the years (Haynes et al. 2005) including electrochemically roughened electrode surfaces (Stacy and Duyne 1983), metal colloidal monolayers (Fan and Brolo 2008) and metal colloids in which roughening can be achieved by inducing colloid aggregation (Aroca et al. 2005). Metal colloids of silver and gold are the most frequently used due to their localized surface plasmon resonance (LSPR) characteristics (Kvítek and Prucek 2005).

The SERS effect strongly depends on the nature of the SERS-active substrate. Particle size, roughness and type of the metal dictate the LSPR properties, which influence the EMM (Brus 2008). The characteristics of the surface coverage of the nanometallic surface influence the affinity of the molecule to the surface, the type of interaction between the molecule and the surface (physical/chemical interaction) and the orientation of the molecule at the surface. Therefore, the design of the SERS-active substrate for a specific application requires a careful consideration of several factors including the incident light wavelength, the LSPR characteristics of the SERS substrate, the surface coverage, the target molecule's nature and possible interferences.

Several procedures for obtaining SERS active metallic colloids have been reported. The most employed one is based on a chemical reduction technique which usually involves a metal salt, reducing agent and stabilizer as the starting materials (in some of the procedures, the reducing agent may serve also as the stabilizer of the nanoparticles). By changing the reactions conditions, one may influence the LSPR characteristics of the nanoparticles (Aroca et al. 2005; Qin et al. 2006).

In the present paper, the effect of target molecules and SERS-active substrate properties on the obtained SERS, using a field portable Raman spectrometer, is reported. Also reported herein is the SERS detection of the chemical warfare agent sulfur mustard (HD, 2,2 dichloroethyl sulfide). This study may serve as a basis for the development of SERS platform for homeland security purposes.

6.2 Experimental

6.2.1 Chemicals

Silver nitrate, hydroxylamine hydrochloride, trisodium citrate, Hydrogen tetrachloroaurate trihydrate, potassium nitrate, nitric acid, sodium hydroxide, benzoic acid, dibenzyl disulphide, p-mercaptobenzoic acid, sodium sulphate, methanol and ethanol were purchased from Sigma-Aldrich (Milwaukee, WI). All chemicals were used without further purification. Ultrapure deionized water (Barnstead Nanopure system Dobuque, IA) was used throughout the experiments. Silver and gold colloids were prepared according to the literature using trisodium citrate (Lee and Meisel 1982) or hydroxylamine hydrochloride (Brown et al. 2000; Leopold and Lendl 2003) as the reducing agent. Stock solutions of benzoic acid, dibenzyl disulphide, p-mercaptobenzoic acid and HD were prepared in ethanol or methanol.

6.2.2 Raman Measurements

Raman spectra were collected using a FirstDefender™ hand held Raman spectrometer (Ahura Scientific, Inc.). Excitation was provided by a 785 nm diode laser with an output power of 300 mW. Samples were placed in glass vials of 1.46 cm in diameter; accumulation time ranged from 0.25 to 10 s.

6.2.3 TEM/SEM Measurement

The colloids were characterized (size and shape) by FEI Tecnai T-12-120 kV transmission electron microscope and by FEI Quanta – FEG200 (high vac. 10 kV) scanning electron microscope. Colloidal suspension (5 μl) was added as it is or diluted by ultrapure water (1:10) onto formvar/carbon coated copper grid (400 mesh). The sample was dried for 1 min before the measurement.

6.2.4 Sample Aggregation

5 μL of the target molecule solution were added to 1 mL colloid and agitated for 30 s. Aggregation of the colloids was obtained after adding varying amounts of KNO_3, HNO_3 or Na_2SO_4 as aggregating reagents to the mixtures, followed by additional agitation of the obtained solution for 30 s prior to the Raman measurement.

6.3 Results and Discussion

6.3.1 Effect of the Protective Layer (Stabilizer)

Silver colloids prepared by the Lee and Meisel's process (Lee and Meisel 1982) were stabilized by the reducing reagent sodium citrate (a citrate anion protective layer). Utilizing the reducing reagent hydroxylamine hydrochloride (Leopold and Lendl 2003) leads to the formation of colloids that are apparently stabilized by chloride ions. Organic molecules that function as protective layer can improve the affinity of hydrophobic molecules to the surface; however, the Raman spectrum of the stabilizers (e.g. citrate anion) may interfere with the analyte's spectrum. SERS of benzoic acid was previously reported (Fleger et al. 2009; Kwon et al. 1994). Here, SERS of benzoic acid was measured using the above two types of colloids as SERS-active substrates (Fig. 6.1). In the presence of sodium citrate-reduced Ag (33 nm in diameter, estimated by TEM), a strong SERS signal of benzoic acid was obtained (Fig. 6.1a); however, in the presence of hydroxyl amine-reduced Ag (32 nm in diameter,

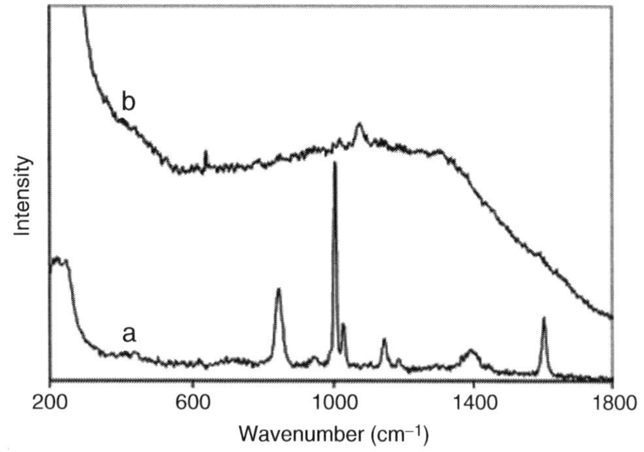

Fig. 6.1 SERS of benzoic acid, $2.5 * 10^{-3}$ M, in the presence of HNO_3, $7.4 * 10^{-4}$ M, and silver colloids: (*a*) sodium citrate-reduced Ag, accumulation time: 250 ms; (*b*) hydroxylamine-reduced Ag, accumulation time: 3 s

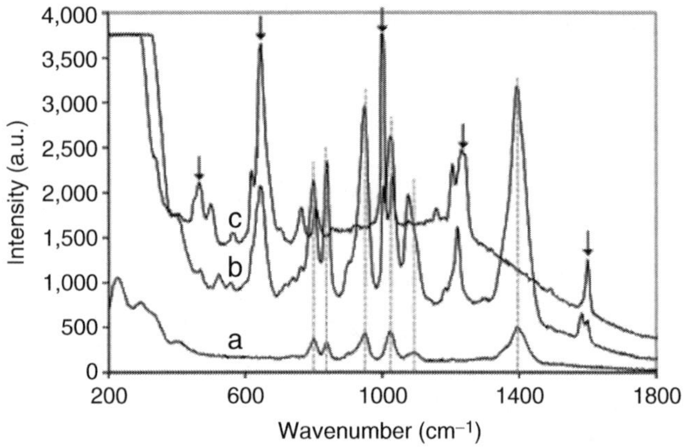

Fig. 6.2 (*a*) SERS of sodium citrate-reduced Ag in the presence of HNO_3, 7.4 * 10^{-4} M; (*b, c*) SERS of dibenzyl disulfide, 2.4 * 10^{-5} M, in the presence of HNO_3, 2.8 * 10^{-3} M, and silver colloids: (*b*) sodium citrate-reduced Ag; (*c*) Hydroxylamine-reduced Ag. Accumulation time: 1 s. The typical bands of dibenzyl disulfide are indicated by *arrows*

estimated by TEM), the SERS signal was hardly seen (Fig. 6.1b). The presence of citrate ion as a protective layer probably increases the affinity of benzoic acid to the surface.

An example of the SERS signal obtained from the protective layer (sodium citrate in the current example) is shown in Fig. 6.2a. Some peaks in this spectrum might interfere with the detection of low concentrations of the analyte (see for example the SERS of dibenzyl disulfide, Fig. 6.2b). Obviously, hydroxylamine-reduced Ag nanoparticles, which have no organic protective layer, produce a clear spectrum of the target molecule, Fig. 6.2c, which may result in an improved LOD values. In the presence of disulfide compounds, the strong affinity of the sulfur to the metallic surface probably overcomes the affinity of the organic part to the metallic surface. Therefore, sers of dibenzyl disulfide was seen in the presence of hydroxylamine-reduced Ag nanoparticles.

6.3.2 Aggregation Effect

Aggregation of metallic colloids results in "hot spot" which significantly increases the SERS effect (Brus 2008). Normally, aggregation is induced by lowering the electrostatic repulsion between the nanoparticles. It should be noted that the analyte itself might also induce aggregation. Figure 6.3a shows the TEM image of sodium citrate-reduced Ag. Figure 6.3b shows that in the presence of benzoic acid (2.5 * 10^{-3} M), without any aggregating reagent, aggregated nanoparticles are observed, as is the case in the presence of benzoic acid and the aggregating reagent HNO_3 (7 * 10^{-4} M), Fig. 6.3c. However, it seems that the extra addition of aggregating reagent produces bigger aggregates.

The choice of aggregating reagent greatly influences the aggregation rate and the size of the obtained aggregates (Yaffe and Blanch 2008). Optimization for the required amount of the aggregating reagent should be done. The SERS of benzoic acid varied observably with type of aggregating reagent (Fig. 6.4). The comparison was done under optimized conditions for each aggregating reagent. Among the three aggregating reagent, nitric acid gave rise to the most intense bands of both the analyte and the protective layer molecules.

6.3.3 Metal-Analyte Interactions

The SERS signal may be affected by several factors related to metal-analyte interactions. One such factor is the nature of analyte adsorption to the metal – chemisorbed molecules are expected to show stronger SERS signals than physisorbed ones. The SERS signal may depend also on the orientation of the

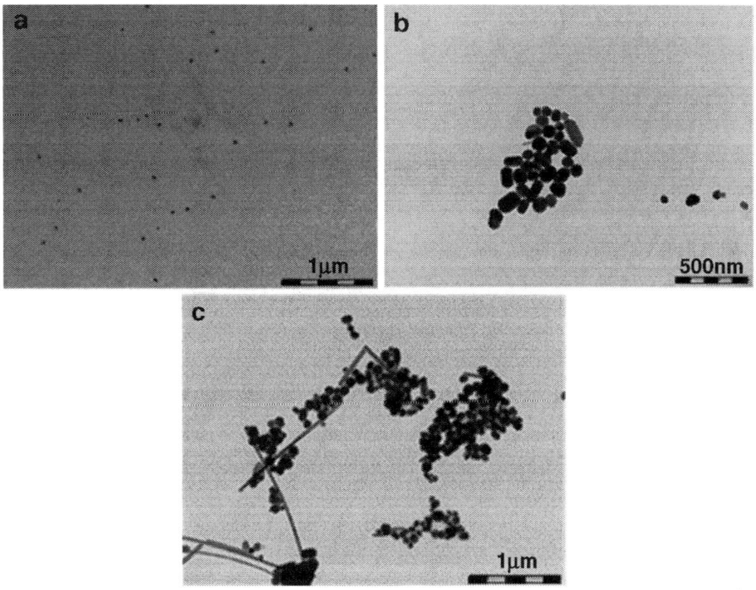

Fig. 6.3 TEM images of sodium citrate-reduced Ag (**a**), in addition of benzoic acid, 2.5 * 10^{-3} M before (**b**) and after (**c**) the addition of HNO_3, 7 * 10^{-4} M

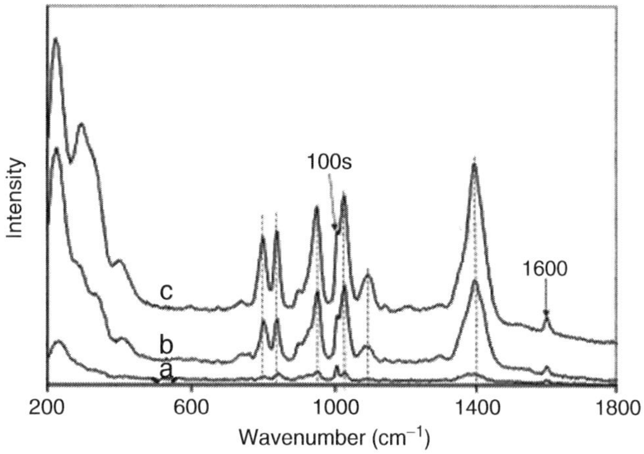

Fig. 6.4 SERS of benzoic acid, 2.5 * 10^{-4} M and aggregation reagents: (*a*) 0.013 M Na_2SO_4; (*b*) 0.038 M KNO_3; (*c*) 3.5 * 10^{-3} M HNO_3. Accumulation time: 250 ms. The typical bands of benzoic acid are indicated by *arrows*; the bands of the protective layer, sodium citrate, are indicated by *dashed lines*

molecule with respect to the metallic surface, which, in turn, may depend on the analyte's concentration. Understanding the effect of analyte concentration on metal-analyte interactions is crucial in order to build a library that will allow us to identify target molecules in a range of concentrations.

For example, p-mercaptobenzoic acid which its SERS was already studied and analyzed in the literature (Fleger et al. 2009; Michota and Bukowska 2003) has several binding sites, the benzene π–electrons, carboxy and thiol groups. Comparison between the spectra at low and high concentrations (Fig. 6.5) shows that at low concentration the weak band at 1,706 cm^{-1} disappeared and the two bands at about 1,387 cm^{-1} and 838 cm^{-1} appear. These changes suggest that at low concentrations the molecules attach to the metal via the COO^- group in addition to the attachment via the thiolate group (1,098 cm^{-1}). At high concentrations, the appearance of the weak band of the C=O group and the disappearance of the COO^-

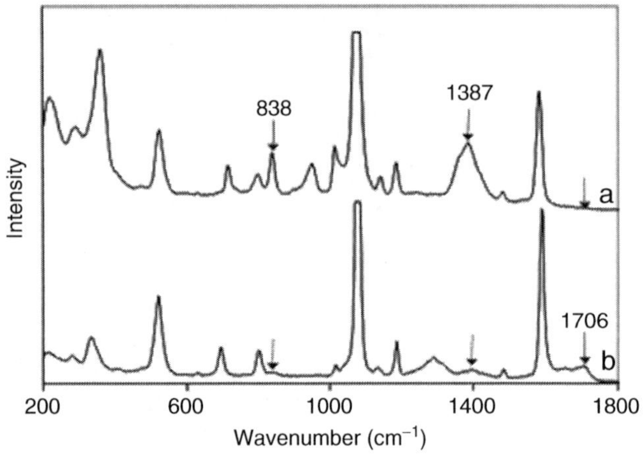

Fig. 6.5 SERS of p-mercaptobenzoic acid in two different concentrations in the presence of sodium citrate-reduced Ag and HNO_3, 2.8 * 10^{-3} M: (*a*) 2.3 * 10^{-6} M, accumulation time: 250 ms; (*b*) 9.4 * 10^{-6} M, accumulation time: 1 s. The typical bands of carboxylic acid and carboxilate group are indicated by *arrows*

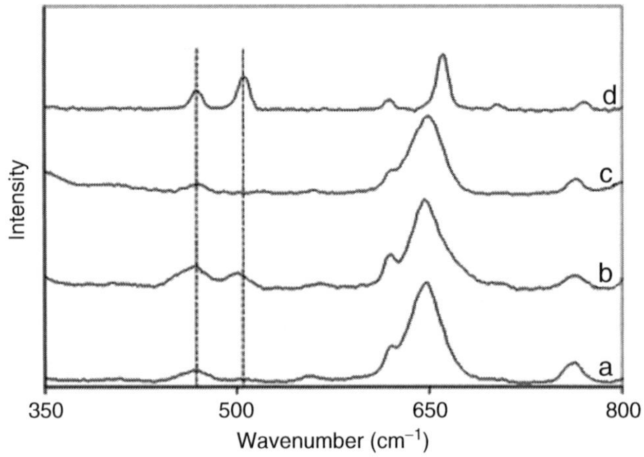

Fig. 6.6 SERS of dibenzyl disulfide, 4.9 * 10^{-5} M, HNO_3, 7.4 * 10^{-4} M, and newly prepared hydroxylamine reduced Ag (*a*) and after 3 days (*b*); (*c*) SERS of dibenzyl disulfide, 4.7 * 10^{-5} M, HNO_3, 2.8 * 10^{-3} M and sodium-citrate reduced Ag. Accumulation time: 250 ms; (*d*) normal Raman of dibenzyl disulfide in the solid state, accumulation time: 1 s

bands suggest that the COOH groups are hydrogen bonded and the attachment is via the thiolate group. Based on the behavior of this molecule as a dependence of concentration, we added SERS spectra to the instrument's library in a range of concentrations that covers the expectable changes. LOD values of p-mercaptobenzoic acid were determined based on the ability to get an identification of the analyte by the instrument. The LOD concentrations that were measured are: 1.2 * 10^{-8} M and 1.2 * 10^{-6} M using sodium-citrate-reduced Ag and hydroxylamine-reduced Ag, respectively. We also determine the enhancement factor (EF) of the SERS effect by comparing the normalized intensity of the SERS spectra to the normalized intensity of the normal Raman of p-mercaptobenzoic acid. The EF values are 1.2 * 10^7 and 5.7 * 10^6, respectively.

The best characterized attachment of disulfides compounds onto silver surface is via the cleavage of an S-S bond and the attachment of the thiolate group onto the surface (Szafranski et al. 1998). Proof for this cleavage is found in the absence of the typical stretching band of the S-S bond. By comparing the normal Raman spectra of dibenzyl disulphide and benzyl mercaptan in the solid state it was seen that the band at 505.8 cm^{-1} disappears and therefore was assigned to the S-S bond. Colloid aging effect was observed in the presence of hydroxylamine-reduced Ag. This aging effect was affecting the SERS of dibenzyl disulfide. In newly prepared hydroxylamine-reduced Ag, the band at 505.8 cm^{-1} was absent, which means that the attachment of dibenzyl disulfide is via the thiolate group (Fig. 6.6a). Repeating this measurement with the hydroxylamine-reduced Ag particles that were

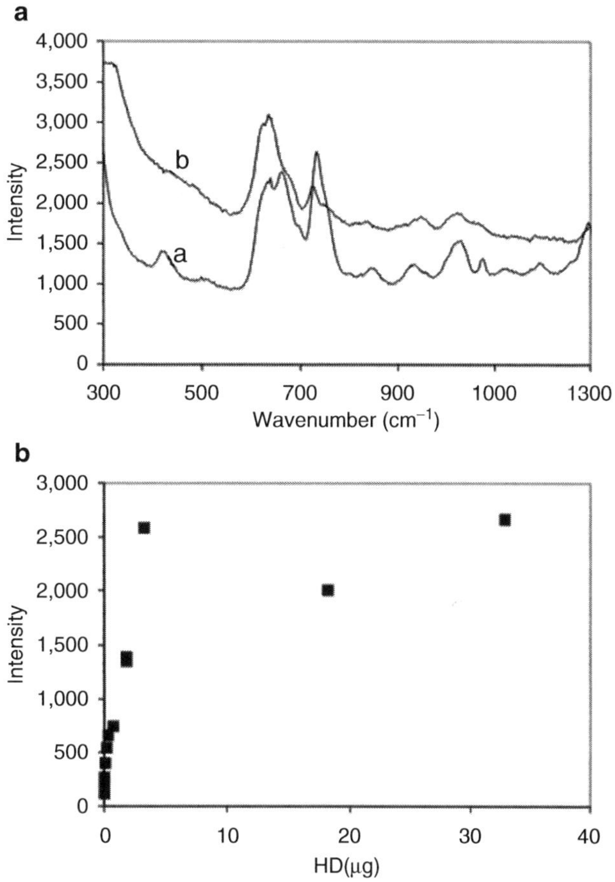

Fig. 6.7 (**A**) SERS of HD in the presence of gold colloids (~60 nm) (*a*) and sodium citrate-reduced Ag (*b*). Accumulation time: 5 s. (**B**) The dependence of SERS spectra of HD on the amount of HD (μg). The intensity was measured at 623 cm^{-1}

stored at 4°C for 3 days resulted in the appearance of a new band at 498.9 cm^{-1}, which was assigned to the S-S stretch (Fig. 6.6b). This attitude was not seen in the case of sodium-citrate-reduced Ag particles (Fig. 6.6c) and it can indicate that the surface of the hydroxylamine-reduced Ag particles is changing with time and possibly their ability to reduce the S-S bond was eliminated. Figure 6.6d shows the typical band of the S-S stretch of normal Raman of dibenzyl disulfide in the solid state. The shift in the frequency of the S-S stretch in the presence of hydroxylamine-reduced Ag particles in comparison to the solid state may be due to the interaction of the molecule with the metallic surface.

6.3.4 Homeland Security Applications

Improvements in portable Raman instruments have increased their applicability to homeland security purposes including the detection and identification of chemical warfare agents (CWA) (Farquharson et al. 2005; Inscore and Farquharson 2005; Inscore et al. 2004). The comparison between the SERS spectra of mustard agent (HD) in the presence of gold nanoparticles and sodium-citrate-reduced Ag particles is shown in Fig. 6.7A. In the presence of gold nanoparticles (60 nm in diameter, estimated by SEM), the major peaks are related to C-S and C-Cl bonds; however in the presence of silver particles, the major peak is related to the C-S bonds and minor peak is related to C-Cl bond. It is assumed that silver nanoparticles enhance the rate of HD hydrolysis resulting in the formation of thiodiglycol, for example. This study is still ongoing. HD LOD is 18 nM (3 ng, S/N = 3) using gold particles and two orders of magnitude higher for silver nanoparticles. Figure 6.7B shows the dependence of SERS spectra of HD on the amount of HD. At low concentrations, the SERS intensity increases linearly with the analyte concentration. As analyte concentration is further increased, the response curve reaches a plateau probably due to surface saturation. At concentrations higher than the maximum value, multi-layers of the adsorbate are formed, rendering SERS effect lower, due to the larger distance between the adsorbate and the surface of the SERS substrate (Dieringer et al. 2006; Kennedy et al. 1999).

6.4 Summary and Conclusions

This work demonstrates the influence of the SERS-active substrates' properties (e.g., protective layer of the nanoparticles), the aggregating reagent and the nature of the target molecule on the properties of the SERS spectra.

Utilizing the SERS effect allows trace detection in the field with a portable Raman instrument. Detection and spectroscopic identification of mercaptobenzoic acid and HD in the nM range is demonstrated.

Since there is no one answer for all target molecules or even for one molecule, SERS substrates should be carefully design according to the target molecules and possible interferences which are typical for on-site applications.

References

Aroca, R. F., Alvarez-Puebla, R. A., Pieczonka, N., Sanchez-Cortez, S., & Garcia-Ramos, J. V. (2005). Surface-enhanced Raman scattering on colloidal nanostructured. *Advanced in Colloid and Interface Science, 116*, 45–61.

Brolo, A. G., Irish, D. E., & Smith, B. D. (1997). Applications of surface enhanced Raman scattering to the study of metal-adsorbate interactions. *Journal of Molecular Structure, 405*, 29–44.

Brown, K. R., Walter, D. G., & Natan, M. J. (2000). Seeding of colloidal Au nanoparticle solutions. 2. Improved control of particle size and shape. *Chemistry of Materials, 12*, 306–313.

Brus, L. (2008). Noble metal nanocrystals: Plasmon electron transfer photochemistry and single-molecule Raman spectroscopy. *Accounts of Chemical Research, 41*, 1742–1749.

Dieringer, J. A., McFarland, A. D., Shah, N. C., Stuart, D. A., Whitney, A. V., Yonzon, C. R., Young, M. A., Zhang, X., & Duyne, R. P. V. (2006). Surface enhanced Raman spectroscopy: New materials, concepts, characterization tools, and applications. *Faraday Discussions, 132*, 9–26.

Dijkstra, R. J., Ariese, F., Gooijer, C., & Brinkman, U. A. T. H. (2005). Raman spectroscopy as a detection method for liquid-separation techniques. *Trends in Analytical Chemistry, 24*, 304–323.

Fan, M., & Brolo, A. G. (2008). Self-assembled Au nanoparticles as substrates for surface-enhanced vibrational spectroscopy: Optimization and electrochemical stability. *ChemPhysChem, 9*, 1899–1907.

Farquharson, S., Gift, A., Maksymiuk, P., & Inscore, F. (2005). Surface-enhanced Raman spectra of VX and its hydrolysis products. *Applied Spectroscopy, 59*, 654–659.

Fleger, Y., Mastai, Y., Rosenbluh, M., & Dressler, D. H. (2009). Surface enhanced Raman spectroscopy of aromatic compounds on silver nanoclusters. *Surface Science, 603*, 788–793.

Haynes, C. L., Yonzon, C. R., Zhang, X., & Duyne, R. P. V. (2005). Surface-enhanced Raman sensors: Early history and the development of sensors for quantitative biowarfare agent and glucose detection. *Journal of Raman Spectroscopy, 36*, 471–484.

Inscore, F., & Farquharson, S. (2005) Detecting hydrolysis products of blister agents in water by surface-enhanced Raman spectroscopy. *Proceedings of the SPIE, 5993*, 19–22.

Inscore, F., Gift, A., Maksymiuk, P., & Farquharson, S. (2004). Characterization of chemical warfare G-agent hydrolysis products by surface-enhanced Raman spectroscopy. *SPIE, 5585*, 46–52.

Kennedy, B. J., Spaeth, S., Dickey, M., & Carron, K. T. (1999). Determination of the distance dependence and experimental effects for modified SERS substrates based on self-assembled monolayers formed using alkanethiols. *The Journal of Physical Chemistry. B, 103*, 3640–3646.

Kvítek, L., & Prucek, R. (2005). The preparation and application of silver nanoparticles. *Journal of Materials Science, 22*, 2461–2473.

Kwon, Y. J., Son, D. H., Ahn, S. J., Kim, M. S., & Kim, K. (1994). Vibrational spectroscopic investigation of benzoic acid adsorbed on silver. *Journal of Physical Chemistry, 98*, 8481–8487.

Lee, P. C., & Meisel, D. (1982). Adsorption and surface-enhanced Raman of dyes on silver and gols sols. *Journal of Physical Chemistry, 86*, 3391–3395.

Leopold, N., & Lendl, B. (2003). A new method for fast preparation of highly surface-enhanced Raman scattering (SERS) active silver colloid at room temperature by reduction of silver nitrate with hydroxylamine hydrochloride. *The Journal of Physical Chemistry. B, 107*, 5723–5727.

Michota, A., & Bukowska, J. (2003). Surface-enhanced Raman scattering (SERS) of 4-mercaptobenzoic acid on silver and gold substrates. *Journal of Raman Spectroscopy, 34*, 21–25.

Moore, D. S., & Scharff, R. J. (2009). Portable Raman explosives detection. *Analytical and Bioanalytical Chemistry, 393*, 1571–1578.

Moskovits, M. (1985). Surface-enhanced spectroscopy. *Reviews of Modern Physics, 57*, 783–826.

Otto, A. (2005). The 'chemical' (electronic) contribution to surface-enhanced Raman scattering. *Journal of Raman Spectroscopy, 36*, 497–509.

Qin, L., Zou, S., Xue, C., Atkinson, A., Schatz, G. C., & Mirkin, C. A. (2006). Designing, fabricating and imaging Raman hot spots. *Proceedings of the National Academy of Science of the United States of America, 103*, 13300–13303.

Smith, W. E. (2008). Practical understanding and use of surface enhanced Raman scattering/surface enhanced resonance Raman scattering in chemical and biological analysis. *Chemical Society Reviews, 37*, 955–964.

Stacy, A. M., & Duyne, R. P. V. (1983). Surface enhanced Raman and resonance Raman spectroscopy in a non-aqueous electrochemical environment: Tris (2,2′-bipyridine)ruthenium(II) adsorbed on silver from acetonitril. *Chemical Physics Letters, 102*, 365–370.

Szafranski, C. A., Tanner, W., Laibinis, P. E., & Garrell, R. L. (1998). Surface-enhanced Raman spectroscopy of aromatic thiols and disulfides on gold electrodes. *Langmuir, 14*, 3570–3579.

Yaffe, N. R., & Blanch, E. W. (2008). Effect and anomalies that can occur in SERS spectra of biological molecules when using a wide range of aggregating agents for hydroxylamine-reduced and citrate-reduced silver colloids. *Vibrational Spectroscopy, 48*, 196–201.

Chapter 7
Quantum Dots and Fluorescent Protein FRET-Based Biosensors

Kelly Boeneman, James B. Delehanty, Kimihiro Susumu, Michael H. Stewart, Jeffrey R. Deschamps, and Igor L. Medintz

Abstract There has been considerable recent interest in the creation of nanoparticle-biomolecule hybrid materials for uses such as *in vitro* and *in vivo* biosensing, biological imaging, and drug delivery. Nanoparticles have a high surface to volume ratio, making them capable of being decorated with various biomolecules on their surface which retain their biological activity. Techniques to bind these biomolecules to nanoparticle surfaces are also advancing rapidly. Here we demonstrate hybrid materials assembled around CdSe/ZnS core/shell semiconductor quantum dots (QDs). These intrinsically fluorescent materials are conjugated to the fluorescent proteins YFP, mCherry and the light harvesting complex b-phycoerythrin (b-PE). QDs have fluorescent properties that make them ideal as donor fluorophores for Förster resonance energy transfer (FRET) while the fluorescent proteins are able to act as FRET acceptors displaying many advantages over organic dyes. We examine FRET interactions between QDs and all three fluorescent proteins. Furthermore, we show QD-mCherry hybrid materials can be utilized for *in vitro* biosensing of caspase-3 enzymatic activity. We further show that QDs and fluorescent proteins can be conjugated together intracellularly with strong potential for live-cell imaging and biosensing applications.

Keywords Biosensing • Cell imaging • Fluorescent proteins • Förster resonance energy transfer (FRET) • Quantum dots (QDs) • Semiconductor • Fluorescence • Enzyme • Caspase

7.1 Introduction

Luminescent semiconductor quantum dots (QDs) have many advantages over commercial dyes, including superior photophysical properties such as high quantum yields, resistance to photobleaching and a large, 'effective' Stokes shift. Because of these properties, they have been used extensively for biological labeling, imaging and sensing applications (Lidke et al. 2004; Alivisatos et al. 2005;

K. Boeneman • J.B. Delehanty • J.R. Deschamps • I.L. Medintz (✉)
Center for Biomolecular Science and Engineering, Code 6900, 4555 Overlook Avenue SW, Washington, DC 20375, USA
e-mail: igor.medintz@nrl.navy.mil

K. Susumu • M.H. Stewart
Optical Sciences Division, Code 5611, US Naval Research Laboratory, Washington, DC 20375, USA

E. Zahavy et al. (eds.), *Nano-Biotechnology for Biomedical and Diagnostic Research*, Advances in Experimental Medicine and Biology 733, DOI 10.1007/978-94-007-2555-3_7,

Medintz et al. 2005; Michalet et al. 2005; Klostranec and Chan 2006; Delehanty et al. 2009a, b). Many techniques to conjugate biomolecules to the QD surface have been developed, for example covalent-chemical surface attachment and metal affinity coordination, (Frasco and Chaniotakis 2010; Li et al. 2010; Sperling and Parak 2010; Medintz et al. 2005) and numerous examples of QD-based biosensors can be found in the current literature. Many of these biosensors exploit fluorescence resonance energy transfer (FRET) between a QD donor and an organic dye acceptor bound to a biomolecule for signal transduction. Construction of these hybrid nanomaterials usually involves labeling of the biomolecule, as well as the subsequent conjugation reaction of the biomolecule to the QD surface.

Here, in contrast to labeling the biomolecules with an organic dye, we use intrinsically fluorescent proteins. The most significant example, Green fluorescent protein (GFP) which is isolated from the jellyfish *Aequorea victoria*, has an intrinsic fluorophore that matures upon proper folding allowing it to fluoresce green under UV light. It has been used for decades for protein labeling and imaging via GFP-protein fusions. Yellow and cyan variants of GFP (YFP and CFP) have been created in the laboratory to allow for multicolor imaging and are now commercially available. The discovery of DsRed, isolated from *Discosoma* sp. reef coral, opened up even more color opportunities in the red spectrum. Use of DsRed was initially limited as it had a long maturation time and fluoresced only as a tetramer, but Roger Tsien's group created a line of monomeric DsRed derivatives with a range of emission wavelengths (Shu et al. 2006). Today, these and other, fluorescent proteins are available in an array of colors and with superior fluorescent properties compared to their predecessors.

There is little currently available in the literature concerning utilizing these new fluorescent proteins in conjunction with QDs. By incorporating a naturally fluorescent protein as a QD FRET acceptor instead of a dye-labeled biomolecule, the dye-labeling step is eliminated. Furthermore, because these are proteins that can be expressed and purified in the laboratory, they can be engineered via common molecular biology techniques to offer an array of functions for biosensing and imaging applications. Here, we utilize two of these fluorescent proteins, YFP and mCherry, along with the light-harvesting complex b-phycoerythrin (b-PE) for conjugation to CdSe/ZnS QDs. We demonstrate FRET between the QDs and all three fluorophores. Furthermore, we demonstrate how a QD-mCherry construct can be engineered to be a nanoscale biosensor for caspase-3 activity. Lastly, we show that QDs and mCherry can be conjugated *in vivo* within cell culture, adding the possibility that QD-fluorescent protein hybrid materials may potentially be directly annealed *in situ* and then used as intracellular labeling and biosensing agents.

7.2 Materials and Methods

7.2.1 Quantum Dots

CdSe/ZnS core-shell QDs with the indicated emission maxima were synthesized using a high temperature reaction of organometallic precursors in hot coordinating solvents (Clapp et al. 2006). Hydrophilicity was obtained by cap exchange with either dihydrolipoic acid (DHLA), DHLA-PEG600 or biotin-terminated DHLA-PEG400 (9:1 ratio DHLA-PEG600:DHLA400-biotin) (Susumu et al. 2007). Invitrogen's Qdot 565 nm ITK carboxyl QDs were used for microinjection experiments.

7.2.2 Fluorescent Proteins

Both YFP and mCherry were obtained in pRSET B plasmids (Invitrogen, Carlsbad, CA) and expressed as described (Medintz et al. 2009; Fehr et al. 2002; Medintz et al. 2008). Maltose binding protein was purified as described previously (Clapp et al. 2004). 1 mg/ml b-PE-streptavidin conjugates were obtained from Invitrogen. The caspase-3 recognition sequence's DEVD and SGDEVDSG were added to the mCherry expression vector using Stratagene's QuickChange site-directed mutagenesis kit following the manufacturer's recommended protocol (Agilent Technologies, Santa Clara, CA) as described in Boeneman et al. (2009). For live-cell experiments, the mCherry coding sequence in the pmCherry N1 eukaryotic expression vector (Clontech) was replaced with His_6-mCherry. The resulting plasmid was confirmed by restriction digestion and DNA sequencing. The parent pmCherry-N1 vector and the modified His_6-pmCherry N1 vector were transfected into COS-1 cells using Lipofectamine 2000 (Invitrogen, Carlsbad,CA) as described in (Boeneman et al. 2010). mCherry expression was observed in approximately 40% of transfected COS-1 cells.

7.2.3 Self-Assembly of Quantum Dot Bioconjugates

Hexahistidine-tagged YFP and mCherry were conjugated to DHLA coated QDs via metal affinity coordination (Sapsford et al. 2007). Assembly of His_6-proteins to ITK QDs is as described in (Boeneman et al. 2010). Valence was controlled by the number of proteins per QD in the reaction mixture. His-labeled YFP was conjugated in varying ratios with His-labeled MBP to maintain a constant valence of 12 proteins per QD. This controlled for QD PL issues associated with varying number of His-proteins on the QD surface (Medintz et al. 2009). YFP, MBP and mCherry were all self-assembled onto DHLA QDs for 30 min as previously described (Medintz et al. 2003). b-PE-streptavidin was conjugated with DHLA-PEG-biotin QDs overnight at 4°C in phosphate buffered saline (PBS) pH 7.4.

7.2.4 Steady-State FRET Data Collection and Analysis

Steady-state fluorescent spectra from solutions of QD-protein bioconjugates and acceptor proteins alone (controls) were collected on a Tecan Safire Dual Monochromator Multifunction Microtiter Plate Reader (Tecan, Research Triangle Park, NC) using 325 nm excitation. Following subtraction of the direct excitation contribution to the acceptor from control solutions, composite spectra were deconvoluted to separate QD and fluorescent protein signal. Energy transfer efficiency E was extracted for each set of QD-conjugates using the expression:

$$E = \left(F_D - F_{DA}\right) / F_D \tag{7.1}$$

where F_D and F_{DA} are, respectively, the fluorescence intensities of the donor alone and donor in the presence of acceptor(s). The metal-His driven self-assembly (applied to YFP and mCherry) provides conjugates with a centro-symmetric distribution of acceptors around the QD. This is assumed to result in a constant average center-to-center separation distance r (Clapp et al. 2004; Medintz et al. 2006). If analyzed within the Förster dipole-dipole formalism, energy transfer efficiency can be fit to: (Clapp et al. 2004):

$$E = nR_0^{\ 6} / \left(nR_0^{\ 6} + r^6\right) \tag{7.2}$$

where n is the average number of fluorescent proteins per QD, and R_0 is the Förster separation distance corresponding to 50% energy transfer efficiency (Lakowicz 2006). For conjugates having small numbers of acceptors ($n<5$), heterogeneity in the conjugate valence was taken into account using the Poisson distribution function, $p(N,n)$, when fitting the efficiency data (Pons et al. 2006):

$$E = \Sigma p\left(N,n\right)E(n)\, and\, p\left(N,n\right) = N^n e^{-N} / N! \tag{7.3}$$

where n designates the exact numbers of acceptors (valence) for conjugates with a nominal average valence of N. We have previously shown that the valence of QD-protein bioconjugates formed, for example, via metal-affinity self-assembly follows a statistical distribution that is well described by this Poisson function, see Pons et al. (2006) for additional details.

7.2.5 *Measurement of Caspase-3 Activity*

Calibration curves were created to equate QD and mCherry PL ratios with the number of mCherry proteins attached to the QD surface, as well as to choose an optimal protein to QD valence to show an efficient change in FRET signal upon cleavage. Calibration details and the caspase-3 assay set up protocol are described in (Boeneman et al. 2009). Briefly, mCherry-substrate proteins were self-assembled to DHLA QDs in 0.4X PBS pH 8.0 at room temperature for about 15 min. Reactions were aliquoted into 0.5 ml PCR tubes at concentrations ranging from 7.5 to 50 pmols of QD per reaction. 65 units of Caspase-3 (Calbiochem, San Diego, CA) were added per 100 μl of reaction. One sample containing no caspase was used as a negative control. Samples were incubated for 30 min at 30°C. The reactions were halted with 5 μl of a 7.5 mg/ml stock of α-iodoacetamide in 0.4X PBS pH 8.0. Samples were transferred to a 96-well plate and emission spectra were read on the Tecan Plate Reader.

The ratio of donor: acceptor PL at 610 nm and QD PL at 550 nm is used to determine the average number of mCherry/QD in the negative control by direct comparison to the calibration curve above. For each assay point, the same process indicates the remaining ratio of mCherry/QD following proteolysis and by comparison to the negative control, the ratio cleaved during proteolysis is derived. This ratio is then multiplied by the total concentration of QD used at each point to determine the total amount of mCherry cleaved per reaction point. The values are then converted into concentration and divided by time to obtain enzymatic velocity in nM mCherry/cleaved per min. Fitting data with the Michaelis-Menten equation allows us to estimate the corresponding Michaelis constants K_M and maximal velocity V_{max} using:

$$V_0 = d[P]/dt = V_{\max}[S]/(K_M + [S]) \tag{7.4}$$

where [S] is substrate concentration, [P] is product (cleaved peptide), and *t* is time.

7.2.6 *Cell Cultures, Microinjection and Microscopy*

A complete protocol for culturing the COS-1 cells and the preparation and microinjection of ITK QDs can be found in Boeneman et al. (2010). Femtoliter aliquots of ITK QDs supplemented with and without Ni^{2+} were directly injected into the mCherry transfected adherent COS-1 cells using an InjectMan® NI2 micromanipulator equipped with a FemtoJet programmable microinjector (Eppendorf, Westbury, NY). Epifluorescence image collection was carried out using an Olympus IX-71 microscope where samples were illuminated with a Xe lamp for UV excitation or a visible/bright light source. Differential interference contrast images (DIC) were collected using a bright light source.

7.3 Results

7.3.1 *Steady-State FRET*

Steady state FRET experiments were performed between DHLA-QDs and mCherry and YFP and DHLA-PEG/DHLA-PEG-biotin QDs and b-PE. The photophysical properties of the QDs and the proteins are listed in Table 7.1 and the absorption and emission spectra along with the spectral overlaps of the FRET pairs are shown in Fig. 7.1.

Table 7.1 Select quantum dot and fluorescent protein properties

Quantum dot donor			Fluorescent protein acceptor						
QD λ_{max} emission (nm)	Quantum yield (%)	Conjugation strategy	Protein acceptor	λ_{max} absorption (nm)	λ_{max} emission (nm)	Extinction coeff. (M^{-1} cm^{-1})	Quantum yield (%)	R_0 (Å)	r (Å)
510	12	Metal-affinity	**YFP**	516	529	20,200	60	39	62
550	20	Metal-affinity	**mCherry**	587	610	71,000	22	49	56
520	20	Biotin/streptavidin	**b-PE**	545,565	575	2,410,000	98	44[a]	–
540	19	Biotin/streptavidin	**b-PE**	545,565	575	2.410,000	98	53[a]	–

Standard deviations for the experimental r values were <10%. (Medintz et al. 2009)
YFP yellow fluorescent protein, *b-PE* b-phycoerythrin
[a]R_0 calculated for interactions with a single bilin chromophore

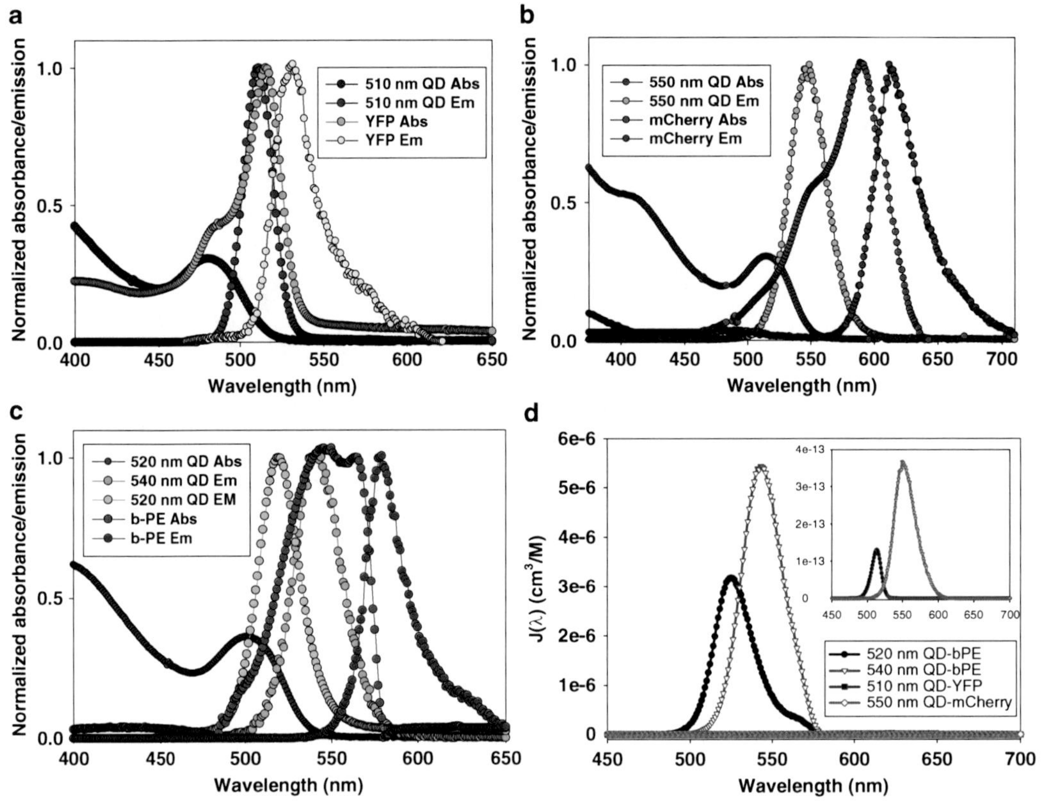

Fig. 7.1 (**a–c**) Absorption and emission spectra of 510 nm QD/YFP, 550 nm QD/mCherry and 520 and 540 nm/b-PE FRET pairs, respectively. (**d**) Spectral overlap functions of the FRET pairs shown in **c** (Adapted from Medintz et al. 2009)

All bioconjugates were excited at 325 nm, to excite the QDs and allow for minimal excitation of the fluorescent protein acceptors. Some direct excitation of the acceptors, however, did occur and was subtracted from the final data. Figure 7.2 shows representative deconvoluted data (a–c) and calculated FRET efficiencies (d–f) for each FRET pair. In all cases, a decrease in QD PL is observed with increasing ratios of fluorescent protein. Acceptor sensitization is also observed and acceptor sensitized emission efficiency is also plotted in (d–f).

While all three donor/acceptor pairs showed FRET activity, QD quenching and acceptor sensitization varied greatly depending on the photophysical properties of the flurophores and the spectral overlaps between the pairs. Overall, data suggests QDs are able to act as FRET donors for a variety of fluorescent proteins. Fluorescent lifetime data and molecular modeling on these pairs are described in Medintz et al. (2009) and both complement and support this observation.

7.3.2 Caspase-3 Sensor

Caspase-3 is a widely studied protease enzyme that is activated during apoptosis, or programmed cell death. It is of particular interest in cancer research, as the lack of this enzyme can result in tumor formation. These same characteristics make it a potentially important pharmaceutical target. For these reasons, we

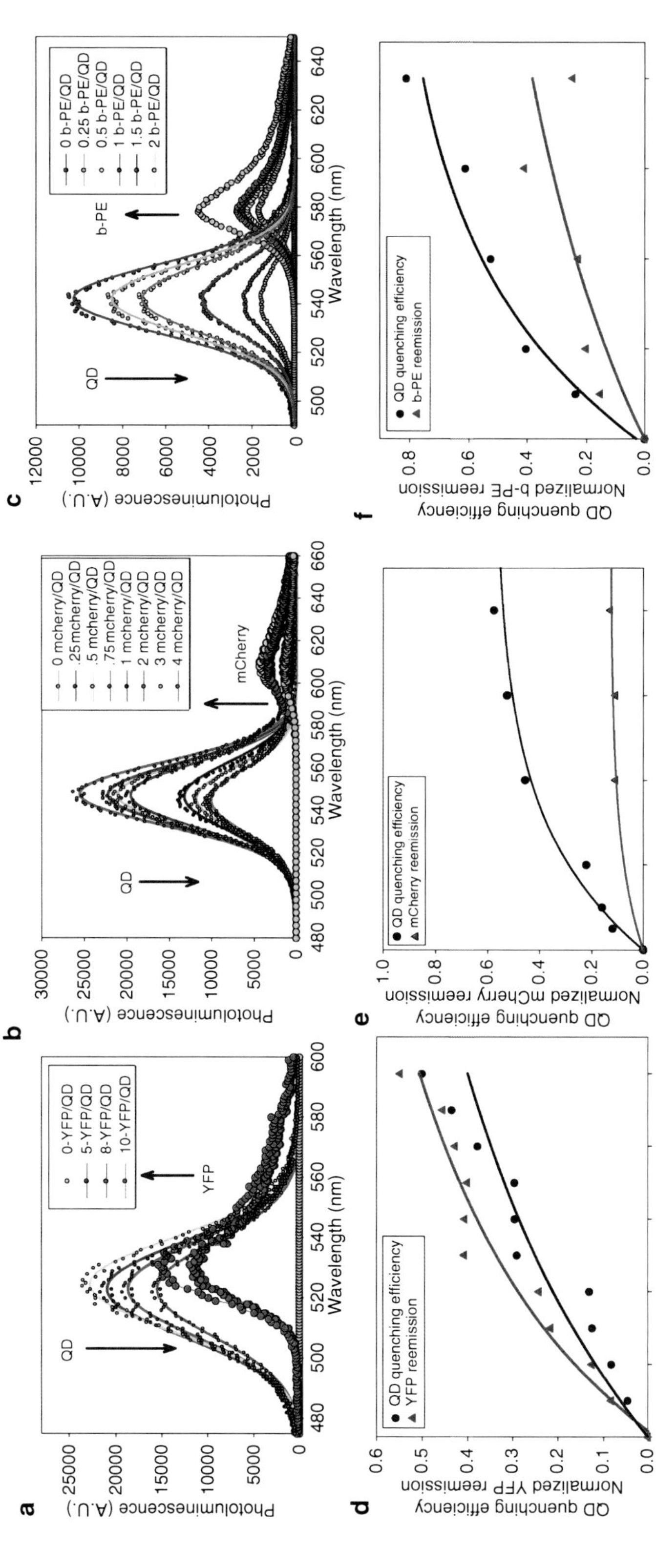

Fig. 7.2 (**a–c**) Representative FRET data for 510 nm QD/YFP, 550 nm QD/mCherry and 540 nm/b-PE FRET pairs respectively. (**d–f**) Calculated FRET efficiency and acceptor sensitization efficiency for **a–c** (Adapted from Medintz et al. 2009)

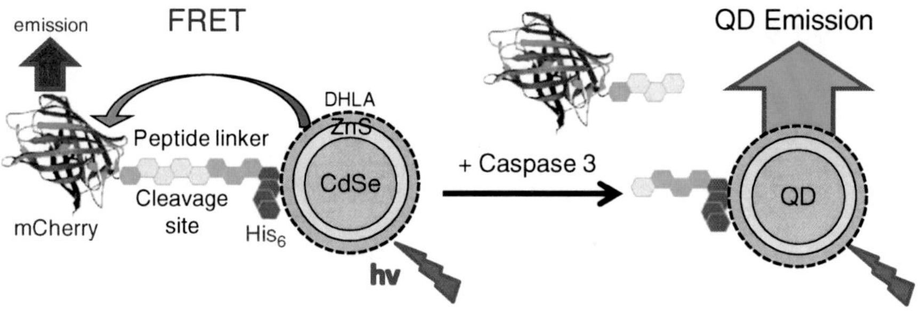

Fig. 7.3 Schematic of QD-mCherry caspase-3 biosensor (Adapted from Boeneman et al. 2010). mCherry with an N-terminal linker expressing the caspase 3 cleavage site and a His_6 sequence were self-assembled to the surface of CdSe-ZnS DHLA QDs, resulting in FRET quenching of the QD and sensitized emission from the mCherry acceptor. Caspase 3 cleaves the linker, reducing the FRET efficiency

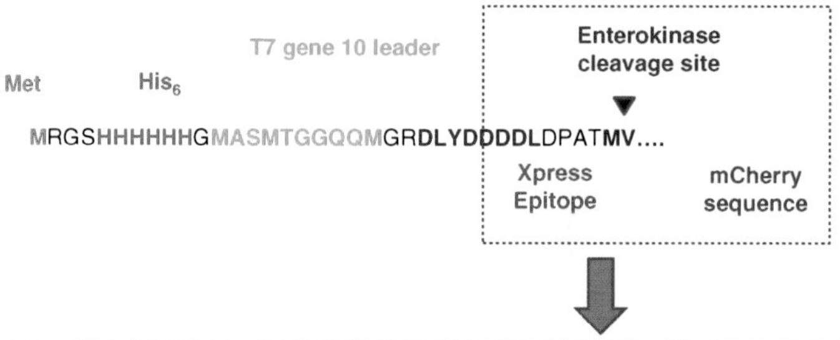

	Peptide Sequence
Original sequence	DLYDDDDLDPATMV...
DEVD insert	DEVDLYDDDDLDPATMV...
SGDEVDSG insert	DSGDEVDSGLYDDDDLDPATMV...

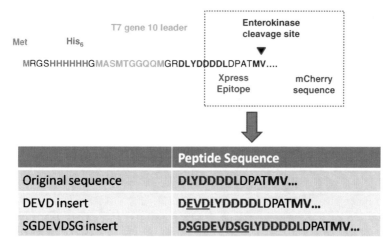

Fig. 7.4 Engineered caspase-3 cleavage sites in mCherry protein. Sites were altered via PCR-based site-directed mutagenesis (Adapted from Boeneman et al. 2009)

chose caspase-3 as a representative enzyme to test the biosensing ability of QD-fluorescent protein hybrid nanomaterials. Due to the ideal photophysical properties and spectral overlap, we chose 550 nm emitting QDs and mCherry as the FRET pair for the biosensor. A schematic of this configuration is shown above in Fig. 7.3. Upon conjugation, FRET occurs between the QD and the mCherry. A caspase-3 cleavage sequence is engineered into the mCherry linker sequence between the QD and the protein fluorophore. In the presence of caspase-3, the linker is cleaved and the FRET interaction is lost. Using calibration curves, the number of QD-mCherry linkages cleaved can be calculated and caspase-3 activity can be derived.

7.3.2.1 Protein Engineering

To sense caspase-3 activity, DNA encoding the caspase-3 cleavage sequence was inserted into the mCherry coding sequence via standard recombinant DNA methods. The site of insertion was determined by protein modeling of the pRSET B 35 residue linker sequence and a region found to have no defined structure was utilized. Caspase-3 recognizes the peptide sequence DEVD, and adding flanking serine-glycine residues (SGDEVDSG) can sometimes improve recognition. Both sequences were inserted into this gene and confirmed by DNA sequencing (Fig. 7.4). Modified and control mCherry

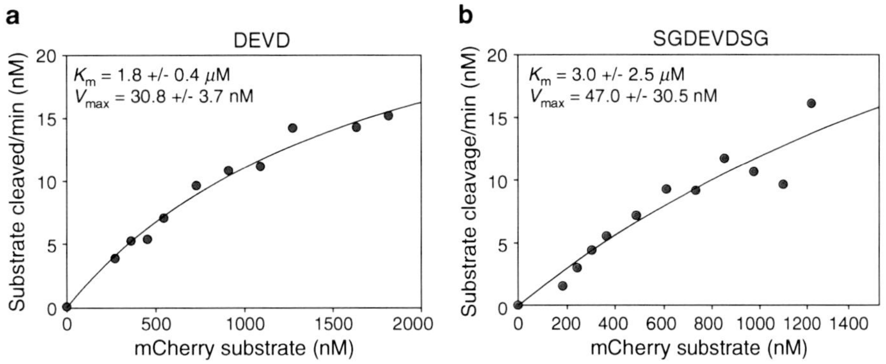

Fig. 7.5 Measured caspase-3 activity mCherry with DEVD and SGDEVDSG cleavage sites. K_m and V_{max} values were derived from the Michaelis-Menten equation (Adapted from Boeneman et al. 2009)

proteins with and without the extended 35 residue linkage sequences were expressed and purified over Ni-NTA resin in the laboratory.

7.3.2.2 Caspase Activity

A calibration curve of QD-mCherry FRET activity at various QD:protein ratios was established prior to the caspase assay. A ratio where a change in FRET activity would be evident was chosen from the calibration curve for the subsequent experiments. The three mCherry protein variants (expressing DEVD, SGDEVDSG, or no cleavage site) were self-assembled to the QDs, and then exposed to 65 units of caspase-3 for 30 min. The reactions were stopped and the emission spectra were taken. Prior to caspase-3 exposure there were approximately four mCherry proteins per QD. After the reaction, the approximate number of acceptors remaining was determined based on comparison to the calibration curve. By knowing the number of QD-mCherry linkages cleaved and the time (30 min), we were able to calculate a cleavage velocity, and thus K_m and V_{max} values using the Michaelis-Menten equation, see Fig. 7.5. The mCherry proteins with both the DEVD and SGDEVDSG sequences showed cleavage activity similar to published caspase-3 enzyme kinetics. The parental mCherry showed no activity.

When compared to established, commercial caspase-3 activity assays, this sensor utilized 5–10 less substrate and three orders of magnitude less enzyme. We were able to detect picomolar levels of caspase-3, making this biosensor extremely sensitive, as well as specific. The caspase-3 recognition sequence can also be easily changed to a recognition sequence for any other protease of interest, making this a versatile biosensing tool with many potential sensing applications.

7.3.3 *Intracellular Delivery and Conjugation*

This above biosensor is extremely useful for *in vitro* detection of caspase-3 and other enzymes; however, ultimate interest is in detection of biological processes inside live cells. The DHLA-QDs utilized have sub-optimal solubility at physiological pH and DHLA-PEG-QDs often hinders the conjugation of large proteins to the QDs due to steric hindrance. Therefore, for cellular studies, we needed a QD that could both bind to the mCherry protein and remain soluble at a cellular pH. Dennis and Bao previously published that EviTag QDs from Evident Technologies, which contain a polymer coating, are

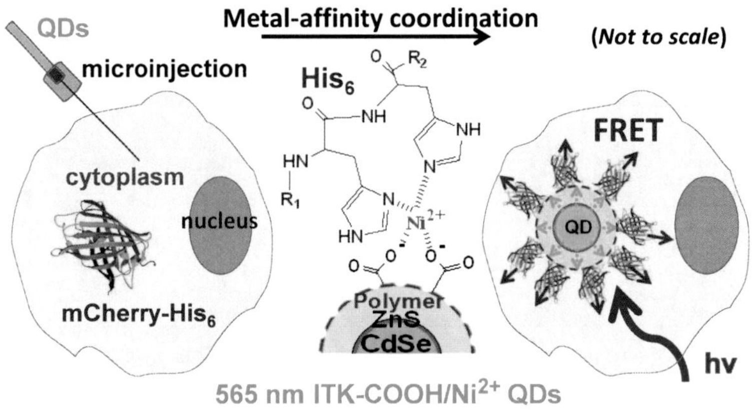

Fig. 7.6 Schematic of intracellular delivery and conjugation of ITK QDs and mCherry. An mCherry expressing vector was transfected into COS-1 cells, allowing mCherry protein to be expressed. Nickel-conjugated QDs (*center*) were microinjected into the cells. Intracellular conjugation was detected by the resulting FRET signal (Adapted from Boeneman et al. 2010)

able to conjugate to mCherry and produce a FRET signal at physiological pH (Dennis and Bao 2008). We screened a number of commercially available QDs for similar properties. We found that ITK carboxyl QDs from Invitrogen, which contained a polymer coating similar to those used by Dennis and Bao, produced the desired results *in vitro*. The free COOH groups were chelated with Ni^{2+}, allowing them to subsequently interact with the hexahistidine tags on the mCherry protein in the cellular cytosol (see Fig. 7.6).

The schematic for the intracellular conjugation process is shown in Fig. 7.6. First, the mCherry gene was cloned into a eukaryotic expression vector and expressed in COS-1 green monkey kidney cells. Then, Ni^{2+} treated QDs were injected into mCherry expressing cells. Separate QD and mCherry expression can be observed in their respective fluorescent channels, see Fig. 7.7. Only upon QD-mCherry conjugation can a yellow FRET signal, originating from QD sensitization of the mCherry acceptor, be observed. Furthermore, QD injection into cells expressing mCherry without a histidine tag did not result in a FRET signal.

This shows that fluorescent and other proteins are able to conjugate to QD surfaces *in vivo* similar to the *in vitro* experiments shown above. While QD microinjection is cumbersome, newer peptide based cellular delivery systems are emerging (Delehanty et al. 2006, 2009, 2010). Quantitative detection of changes in intracellular FRET signal, such as what is detected in the *in vitro* caspase-3 assay above, is difficult and requires specialized equipment for *in vivo* measurements. However, technology is advancing rapidly. The use of QDs to detect cellular processes and signal transduction cascades in live cells is a future possibility, given the growing development and maturity of such nanoparticle sensors.

7.4 Discussion

Fluorescent proteins make ideal FRET acceptors with CdSe/ZnS semiconductor QD donors. These nanobiomolecule hybrids can act as biosensors *in vitro*, and potentially *in vivo*. The myriad of emission wavelengths available in both QDs and fluorescent proteins, as well as the ease of genetic manipulation of the proteins, opens up possibilities for any number of QD/protein combinations and applications. Advances in QD cellular delivery techniques will allow for even more development of

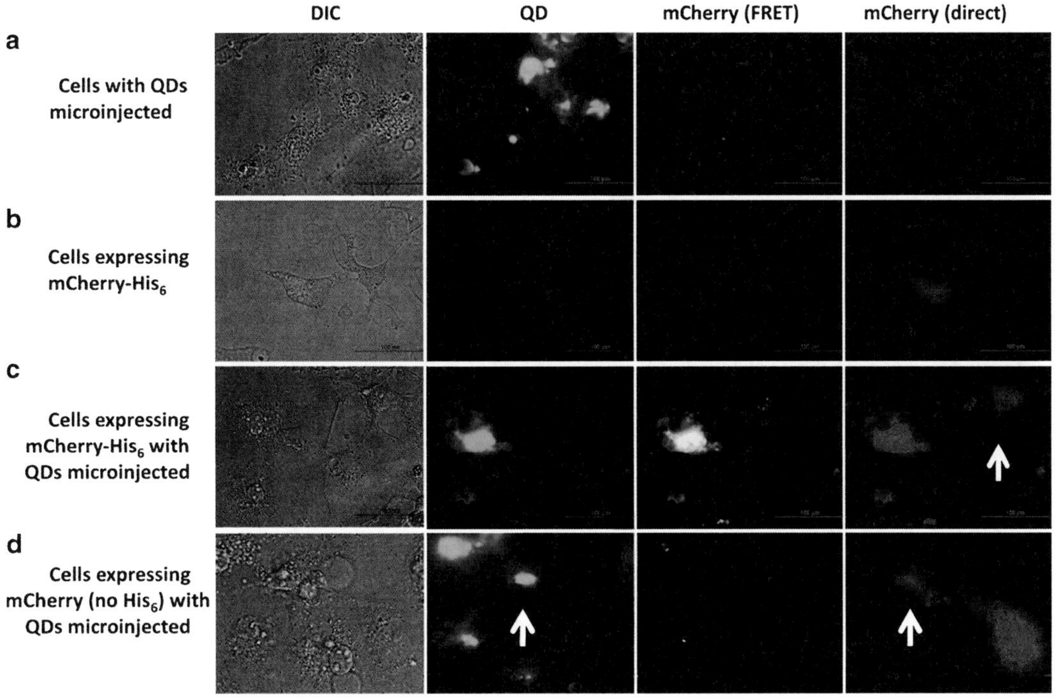

Fig. 7.7 Micrographs showing QD, FRET and mCherry fluorescent channels for cells expressing QD or mCherry alone and together with and without a His-tag. A FRET signal is only detected when QDs and His-tagged mCherry are present together inside the cell, indicative of bioconjugation (Adapted from Boeneman et al. 2010). The arrow in Panel C indicates a cell with mCherry expression, but without QDs. Arrows in Panel D indicate cells with both QDs and mCherry, but no FRET

this field. Therapeutic and diagnostic applications may also begin to emerge (Delehanty et al. 2009b). Technologies perfected in QDs may also be relevant to other types of nanoparticles. Overall, the conjugation of QDs and fluorescent proteins is an important step in the advancement of nanobiotechnology.

Acknowledgements The authors acknowledge the NRL-NSI, ONR, and DTRA.

References

Alivisatos, A. P., Gu, W., et al. (2005). Quantum dots as cellular probes. *Annual Review of Biomedical Engineering, 7*, 55–76.

Boeneman, K., Mei, B. C., et al. (2009). Sensing caspase 3 activity with quantum dot-fluorescent protein assemblies. *Journal of the American Chemical Society, 131*(11), 3828–3829.

Boeneman, K., Delehanty, J. B., et al. (2010). Intracellular bioconjugation of targeted proteins with semiconductor quantum dots. *Journal of the American Chemical Society, 132*(17), 5975–5977.

Clapp, A. R., Medintz, I. L., et al. (2004). Fluorescence resonance energy transfer between quantum dot donors and dye-labeled protein acceptors. *Journal of the American Chemical Society, 126*(1), 301–310.

Clapp, A. R., Goldman, E. R., et al. (2006). Capping of CdSe-ZnS quantum dots with DHLA and subsequent conjugation with proteins. *Nature Protocols, 1*(3), 1258–1266.

Delehanty, J. B., Medintz, I. L., et al. (2006). Self-assembled quantum dot-peptide bioconjugates for selective intracellular delivery. *Bioconjugate Chemistry, 17*, 920–927.

Delehanty, J. B., Mattoussi, H., et al. (2009a). Delivering quantum dots into cells: Strategies, progress and remaining issues. *Analytical and Bioanalytical Chemistry, 393*(4), 1091–1105.

Delehanty, J. B., Boeneman, K., et al. (2009b). Quantum dots: A powerful tool for understanding the intricacies of nanoparticle-mediated drug delivery. *Expert Opinion on Drug Delivery, 6*(10), 1091–1112.
Delehanty, J. B., Bradburne, C. E., et al. (2010). Delivering quantum dot-peptide bioconjugates to the cellular cytosol: Escaping from the endolysosomal system. *Integrative Biology, 2*, 265–277.
Dennis, A. M., & Bao, G. (2008). Quantum dot-fluorescent protein pairs as novel fluorescence resonance energy transfer probes. *Nano Letters, 8*(5), 1439–1445.
Fehr, M., Frommer, W. B., et al. (2002). Visualization of maltose uptake in living yeast cells by fluorescent nanosensors. *Proceedings of the National Academy of Sciences of the United States of America, 99*(15), 9846–9851.
Frasco, M. F., & Chaniotakis, N. (2010). Bioconjugated quantum dots as fluorescent probes for bioanalytical applications. *Analytical and Bioanalytical Chemistry, 396*(1), 229–240.
Klostranec, J. M., & Chan, W. C. W. (2006). Quantum dots in biological and biomedical research: Recent progress and present challenges. *Advanced Materials, 18*, 1953–1964.
Lakowicz, J. R. (2006). *Principles of fluorescence spectroscopy*. New York: Springer.
Li, J., Wu, D., et al. (2010). Preparation of quantum dot bioconjugates and their applications in bio-imaging. *Current Pharmaceutical Biotechnology, 11*(6), 662–671.
Lidke, D. S., Nagy, P., et al. (2004). Quantum dot ligands provide new insights into erbB/HER receptor-mediated signal transduction. *Nature Biotechnology, 22*(2), 198–203.
Medintz, I. L., Clapp, A. R., et al. (2003). Self-assembled nanoscale biosensors based on quantum dot FRET donors. *Nature Materials, 2*(9), 630–638.
Medintz, I., Uyeda, H., et al. (2005). Quantum dot bioconjugates for imaging, labeling and sensing. *Nature Materials, 4*, 435–446.
Medintz, I. L., Clapp, A. R., et al. (2006). Proteolytic activity monitored by fluorescence resonance energy transfer through quantum-dot-peptide conjugates. *Nature Materials, 5*(7), 581–589.
Medintz, I. L., Pons, T., et al. (2008). Intracellular delivery of quantum dot-protein cargos mediated by cell penetrating peptides. *Bioconjugate Chemistry, 19*(9), 1785–1795.
Medintz, I. L., Pons, T., et al. (2009). Resonance energy transfer between luminescent quantum dots and diverse fluorescent protein acceptors. *Journal of Physical Chemistry C, 131*, 18552–18561.
Michalet, X., Pinaud, F. F., et al. (2005). Quantum dots for live cells, in vivo imaging, and diagnostics. *Science, 307*, 538–544.
Pons, T., Medintz, I. L., et al. (2006). Solution-phase single quantum dot fluorescence resonant energy transfer sensing. *Journal of the American Chemical Society, 128*, 15324–15331.
Sapsford, K. E., Pons, T., et al. (2007). Kinetics of metal-affinity driven self-assembly between proteins or peptides and CdSe-ZnS quantum dots. *Journal of Physical Chemistry C, 111*, 11528–11538.
Shu, X., Shaner, N. C., et al. (2006). Novel chromophores and buried charges control color in mFruits. *Biochemistry, 45*(32), 9639–9647.
Sperling, R. A., & Parak, W. J. (2010). "Surface modification, functionalization and bioconjugation of colloidal inorganic nanoparticles". *Philosophical Transactions. Series A, Mathematical, Physical, and Engineering Sciences, 368*(1915), 1333–1383.
Susumu, K., Uyeda, H. T., et al. (2007). Enhancing the stability and biological functionalities of quantum dots via compact multifunctional ligands. *Journal of the American Chemical Society, 129*, 13987–13996.

Chapter 8
Semiconductor Quantum Dots as FRET Acceptors for Multiplexed Diagnostics and Molecular Ruler Application

Niko Hildebrandt and Daniel Geißler

Abstract Applications based on Förster resonance energy transfer (FRET) play an important role for the determination of concentrations and distances within nanometer-scale systems *in vitro* and *in vivo* in many fields of biotechnology. Semiconductor nanocrystals (Quantum dots – QDs) possess ideal properties for their application as FRET acceptors when the donors have long excited state lifetimes and when direct excitation of QDs can be efficiently suppressed. Therefore, luminescent terbium complexes (LTCs) with excited state lifetimes of more than 2 ms are ideal FRET donor candidates for QD-acceptors. This chapter will give a short overview of theoretical and practical background of FRET, QDs and LTCs, and present some recent applications of LTC-QD FRET pairs for multiplexed ultra-sensitive *in vitro* diagnostics and nanometer-resolution molecular distance measurements.

Keywords Diagnostics • FRET • Imaging • Quantum dots • Terbium

8.1 Introduction

Many biological and biochemical processes and functions take place on a nanometer length scale and understanding nano-bio-systems is of great importance for a thorough knowledge of fundamental life sciences as well as for many commercial applications in biosensing and diagnostics. In order to analyze those systems in their natural conformation and surroundings, physiology, function and dynamics can be revealed by introducing spectroscopically and/or microscopically accessible chemical labels.

Förster resonance energy transfer (FRET) can be used to measure structural changes or dynamics at distances ranging from approximately 1–20 nm, which corresponds very well to many biomolecular processes. In 1948 Theodor Förster derived a quantum mechanical FRET theory (Förster 1948) for allowed dipole transitions of donor and acceptor molecules of the same kind (homotransfer). However, the radiationless FRET process can also involve forbidden transitions, and theory as well as methods

N. Hildebrandt (✉)
Institut d'Electronique Fondamentale, Université Paris-Sud 11, Orsay, France
e-mail: niko.hildebrandt@u-psud.fr

D. Geißler
Physikalische Chemie, Universität Potsdam, Potsdam, Germany

E. Zahavy et al. (eds.), *Nano-Biotechnology for Biomedical and Diagnostic Research*,
Advances in Experimental Medicine and Biology 733, DOI 10.1007/978-94-007-2555-3_8,

have been widely expanded (with many different donor and acceptor species), which has been described in several comprehensive books, special issues or reviews (ChemPhysChem 2011; Clegg 1996, 2009; Dale and Eisinger 1974; Dale et al. 1979; Förster 1949, 1959, 1965; Fung and Stryer 1978; Geißler and Hildebrandt 2011; Haas et al. 1978; Jares-Erijman and Jovin 2009; Lakowicz 1999; Morrison 1988; RevMolBio 2002; Sapsford et al. 2006; Selvin 2000; Steinberg 1971; Stryer and Haugland 1967; Stryer 1978; Van der Meer et al. 1994; Wu and Brand 1994). Due to the r^{-6} distance dependence between donor and acceptor, FRET is very sensitive to nanoscale changes and has been used for several biochemical applications such as DNA investigations, imaging, protein-protein interactions or immunoassays (Didenko 2001; Fairclough and Cantor 1978; Hofmann et al. 2010; Jares-Erijman and Jovin 2003; Lakowicz and Geddes 1991; Lakowicz 1999; Morrison 1988; Sapsford et al. 2006; Schuler et al. 2005; Selvin 2000; 2002; Steinberg 1971; Stryer 1978; Szollosi et al. 1998; Wu and Brand 1994).

A combination of QD-acceptors with LTC-donors in time-resolved FRET results in significant sensitivity, distance, and multiplexed detection advantages compared to other donor-acceptor pairs (Charbonnière and Hildebrandt 2008). The long-lived luminescence (usually more than 2 ms) of the LTCs allows time-gated detection leading to a nearly complete suppression of short-lived sample auto-fluorescence as well as fluorescence from directly excited QDs. Due to very large overlap integrals, exceptionally large Förster distances (R_0, the donor-acceptor distance where FRET is 50% efficient) of more than 11 nm can be achieved whereas conventional donor-acceptor pairs have much smaller Förster distances, usually below 6 nm (Van der Meer et al. 1994). Using QDs with emission peaks in between or beyond the well separated LTC luminescence bands allows simultaneous FRET from one LTC to many QDs from the green to the NIR spectral range without significant spectroscopic crosstalk (the detection of LTC or QD emission in the QD or LTC channel due to overlap of the respective emission spectra).

In this chapter we will give a short introduction to FRET, QDs and LTCs, in order to point out the specific advantages of the QD-LTC-FRET combination. We will then review two recent applications, where multiplexed LTC to QD FRET was used for highly sensitive bioanalysis. A time- and spectrally-resolved simultaneous measurement of five FRET-sensitized QDs using LTCs as FRET donors within a biotin-streptavidin bioassay will be presented as a representative application for *in vitro* diagnostics. The color-coded, nearly background-free multiplexed homogeneous assay yields sub-picomolar detection limits for all five QDs in a single sample, making the QD-based FRET probes ideal candidates for highly specific and sensitive companion diagnostics. The same biotin-streptavidin system is then used for the determination of different donor-acceptor distances by time-resolved analysis. This second multiplexed approach demonstrates the advantages of LTC to QD FRET for the application as multiplexed nanometer scale spectroscopic ruler over very large molecular distances (>10 nm). The time-resolved FRET nanoprobes are used for size and shape determination of QDs under physiological conditions and should be well suitable for high-resolution multiplexed conformational and functional studies in cell imaging. Another interesting aspect of the LTC to QD FRET experimental data is the insight into energy transfer mechanisms, which seem inauspicious for the original FRET theory but still follow quite well the r^{-6} distance dependence.

8.2 Short Theoretical and Practical Background

8.2.1 FRET – Förster Resonance Energy Transfer

FRET is an r^{-6} donor-acceptor distance dependent energy transfer process. It is well characterized and established in many biochemical applications. One of the most important parameters for FRET application is the so-called Förster radius or Förster distance R_0 (the distance between donor and acceptor where the energy transfer is 50% efficient). This value can be calculated from relatively easily accessible spectral data of donor luminescence and acceptor absorption. With the overlap integral of donor emission and acceptor absorption spectra $J(\lambda)$ in M^{-1} cm^{-1} nm^4, R_0 (in Å) can be calculated by:

$$R_0^6 = 8.79\times 10^{-5}\, n_r^{-4} \Phi_D \kappa^2 J(\lambda) \tag{8.1}$$

n_r: refractive index of the surrounding medium (e.g. $n_r = 1.33$ for water),
Φ_D: luminescence quantum yield of D,
κ^2: orientation factor taking into account the relative orientations of the D and A dipoles.

The FRET efficiency η_{FRET} can be calculated by using distances (R_0 and r), luminescence quantum yields (Φ), decay times (τ) or intensities (I) of D in the absence (subscript "D") and in the presence (subscript "DA") of A:

$$\eta_{FRET} = \frac{R_0^6}{R_0^6 + r^6} = 1 - \frac{\Phi_{DA}}{\Phi_D} = 1 - \frac{\tau_{DA}}{\tau_D} = 1 - \frac{I_{DA}}{I_D} \tag{8.2}$$

Thus, Eq. 8.2 can be used to calculate unknown donor-acceptor distances r by the acquired spectroscopic data. The larger R_0, the larger the possible distances to be measured efficiently or the higher the efficiency at a fixed donor-acceptor distance. To date R_0 values for commonly used donor-acceptor pairs are rarely larger than 6 nm (Van der Meer et al. 1994). The use of QDs enables Förster distances in the 4–7 nm range, whereas the use of lanthanides as donors can lead to R_0 values as high as 9 nm (Charbonnière et al. 2006). Combining both lanthanides (especially LTCs) and QDs as FRET pair, Förster radii of more than 10 nm become accessible (Charbonnière and Hildebrandt 2008). Thus breaking the commonly cited limit of 10 nm is possible, and investigation of QDs and LTCs for biochemical FRET applications becomes extremely valuable. Another important aspect of using LTCs is their long excited state lifetime, which can be larger than 2 ms and is thus several orders of magnitude longer than the one of QDs (in the 10–100 ns range). Therefore, the luminescence decay time of the QD-acceptors in the presence of the LTC-donor (τ_{AD}) can be used in Eq. 8.2 instead of the donor luminescence decay time ($\tau_{AD} = \tau_{DA}$) (Charbonnière and Hildebrandt 2008). This aspect is of paramount importance for multiplexed measurements with one LTC-donor and several different QD-acceptors because the FRET signals can be detected (and distinguished) via the different QD-acceptor wavelengths. This section is supposed to give a very short overview of FRET in order to highlight the advantages of LTC to QD FRET for nano-bio-analysis. For more details about FRET theory and applications, the interested reader is referred to the FRET literature (and references therein) cited in the introduction.

8.2.2 *Quantum Dots (QDs)*

QDs are well-characterized and well-established nanoparticles (especially for CdSe/ZnS core/shell dots) with unique optical and photophysical properties (Alivisatos 1996; Yin and Alivisatos 2005; Murphy and Coffer 2002). They are frequently used in different life science applications (Alivisatos 2004; Chan and Nie 1998; Gill et al. 2008; Katz and Willner 2004; Medintz et al. 2005, 2008; Michalet et al. 2005; Penn et al. 2003; Jamieson et al. 2007; Bruchez et al. 1998; Bruchez 2005; Raymo and Yildiz 2007; Somers et al. 2007) and can be used both as FRET donors (Medintz et al. 2003; Clapp et al. 2004, 2005, 2007; Algar and Krull 2010; Boeneman et al. 2009; Chen et al. 2008) as well as acceptors (Hildebrandt et al. 2005; Charbonnière et al. 2006; Charbonnière and Hildebrandt 2008; Hildebrandt et al. 2007; Hildebrandt and Löhmannsröben 2007; Geißler et al. 2010; Morgner et al. 2010; Härmä et al. 2007; Roberti et al. 2011; Curutchet et al. 2008).

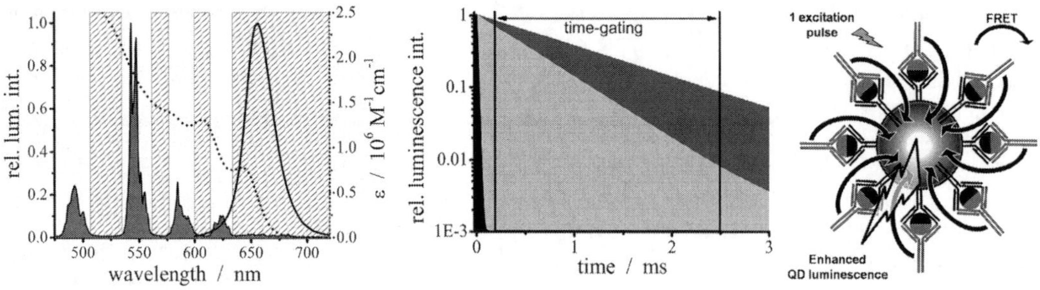

Fig. 8.1 *Left*: Absorption (*dotted line*) and emission spectra of a commercially available QD (Qdot655 – Invitrogen/ Life Technologies) and emission spectrum of a typical LTC. The *shaded areas* in the background represent possible wavelength regions for color multiplexing with FRET from LTCs to QDs. *Middle*: Time-gating results in background (*black*) suppression and detects only the FRET (*light grey*) and pure donor (*dark grey*) luminescence. *Right*: QD labeling with several LTC-conjugated biomolecules (e.g. antibodies – *light grey*) leads to multiple successive FRET processes from the LTCs to the same QD after one excitation pulse, which results in enhanced overall brightness and sensitivity

The main advantages of QDs for bioanalysis are (cf. Fig. 8.1):

- Broad absorption region → Almost no restrictions to (FRET-)excitation wavelengths
- High extinction coefficients and very bright → Highly sensitive detection
- Size tunable absorption and emission wavelengths → Color multiplexing
- Narrow (Gaussian-shaped) emission bands → Almost no spectral overlap for multiplexing
- Very photostable → Longer observation times
- Large surfaces → Labeling with many biomolecules (multiple molecule FRET)
- QDs of different color have equal surface chemistry → One labeling procedure for all

The synthesis of QDs with different sizes is an established technique, and QDs are commercially available (e.g. CAN, eBioscience, Invitrogen/Life Technologies, NN-Labs, PlasmaChem) both in organic solvents (for further treatment in order to yield water solubility and biocompatibility) as well as in aqueous solutions (with functional groups on the surface, as biolabeling kits or with readily labeled biomolecules such as biotin or streptavidin). However, to the best of our knowledge, commercially available diagnostic assays based on QDs do not exist. A mayor drawback for implementation of the available nanoscale QDs into macroscopic FRET applications is related to size and stability. If organic fluorophores are attached via a ligand exchange process, the conjugates have a very close donor-acceptor distance, but the QDs are usually of low long-term stability. An alternative method is surface attachment via covalent linkage to polymer coated QDs, which leads to high colloidal stability on the price of a larger donor-acceptor distance. Moreover, the optical and physical properties of available QDs are found to be not sufficiently reproducible for commercial clinical assays. Still not established are QDs with biofunctional thin – though still stable – coatings, yielding synthetically reproducible homogeneous optical properties and high luminescence quantum yields. Such QDs would not only open the possibility of establishing highly sensitive and flexible "real-life" clinical FRET applications, but also yield QD reference materials, not available to the community today.

8.2.3 Luminescent Terbium Complexes (LTCs)

LTCs can combine the advantageous properties of the inner terbium (Tb) ion and its complexing ligand to be very good FRET donors for biochemical applications (Geißler and Hildebrandt 2011;

Selvin 2002). The ligand fulfills two important requirements, namely shielding of the Tb ion from the surrounding biological medium, and efficient light collection followed by energy transfer to the central ion. The Tb ion itself exhibits very special spectral properties, such as long excited state lifetimes (in the millisecond range) and narrow emission bands. The long lifetimes are especially important for FRET donors in biological applications because they allow for the distinction between short-lived background fluorescence (autofluorescence of the biomolecules and/or buffer as well as directly excited acceptor fluorophores) and the long-lived FRET signal. LTCs and other lanthanide complexes can be synthesized with functional groups for biomolecule conjugation and such complexes are frequently used for biochemical applications (Bünzli 1989, 2004, 2010; Richardson 1982; Selvin 2000, 2002; Bünzli and Piguet 2005; Diamandis 1988; Gudgin Dickson et al. 1995; Hemmilä and Laitala 2005; Yuan and Wang 2005; Sammes and Yahioglu 1996; Hemmilä and Mukkala 2001). LTCs for the use in bioanalysis (e.g. labeled to specific antibodies) are commercially available (e.g. CisBio, Invitrogen/Life Technologies, Lumiphore, PerkinElmer). The generation of efficient FRET to QD-acceptors requires LTCs (or other lanthanide complexes) with long excited state lifetimes as donors (resonance energy transfer without external light excitation, such as BRET (bioluminescence) or CRET (chemiluminescence) is not taken into account). The reason for this is the very efficient excitation of the QDs at all wavelengths below their fluorescence band. Thus, donor and acceptor are always simultaneously excited within the excited state lifetime of the short-lived component. If for example a 1:1 concentrated solution of LTCs and QDs is excited at 337 nm (e.g. nitrogen laser), 1,000 times more QDs (extinction coefficient of ca. $1 \cdot 10^7\ M^{-1} cm^{-1}$ at 337 nm) are promoted to an excited state compared to LTCs (extinction coefficient of ca. $1 \cdot 10^4\ M^{-1} cm^{-1}$ at 337 nm). This situation directly after the excitation makes FRET very inefficient because the energy resonance condition (excited state donor and ground state acceptor) is only fulfilled for the FRET pairs with non-excited QDs. However, due to the long excited state lifetime of the Tb-donors compared to the QD-acceptors, the situation changes completely after some hundreds of nanoseconds. Then, all QDs are back in their energetic ground states whereas the majority of Tb-donors are still in their excited states and FRET becomes very efficient. Using this advantage we have demonstrated very efficient FRET even at a concentration excess of QDs (Geißler et al. 2010; Morgner et al. 2010).

The main advantages of LTCs for QD-based FRET bioanalysis are (cf. Fig. 8.1):

- Very long luminescence decay times (up to ms)
  - Very efficient FRET to QDs becomes possible because the LTC excited state lifetimes are ca. 3–5 orders of magnitude longer.
  - Successive FRET to QDs (due to the big difference in excited state lifetimes, one QD can accept the energy of many LTCs successively) after one single excitation pulse leading to high brightness and high sensitivity FRET systems.
  - Nearly background-free measurements (due to the long luminescence decay times (ms) the QD luminescence (resulting from direct QD excitation) and the biological autofluorescence can be suppressed by time gating – ca. 100 μs after the excitation pulse only LTC and FRET-excited QD luminescence remains).
- Well separated, narrow emission bands over a broad wavelength range.
  - Color multiplexing with FRET from one single LTC to several QDs by measuring the FRET-sensitized QD emission within the wavelength gaps of the LTC emission.
  - Extremely broad overlap of LTC emission with the absorption of many possible QDs leading to very large Förster distances and concomitant high efficiency and sensitivity.
- Large separation between absorption and emission wavelengths
  - Efficient luminescence detection without spectrally overlapping excitation light.

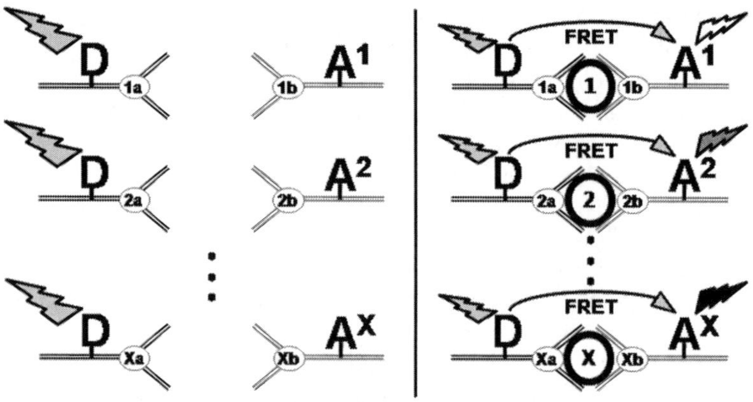

Fig. 8.2 Homogeneous immunoassay principle with several antibodies (1 to X a/b - specific to biomarkers 1 to X) labeled with one donor (D) and several acceptors (A^1 to A^X) for color multiplexing. *Left*: Without biomarkers only D exhibits a long-lived luminescence signal after pulsed light excitation (large D-A distance → no FRET). *Right*: Addition of different biomarkers leads to specific binding and concomitant specific (color-coded) long-lived luminescence signals from A^1 to A^X (small D-A distance → FRET)

8.3 Applications

In this section we review some of our recent results obtained with multiplexed LTC to QD FRET for ultra-sensitive *in vitro* diagnostics (Geißler et al. 2010) and for spectroscopic (or molecular) ruler distance measurements (Morgner et al. 2010) in biological systems. Probably the most important aspects concerning these applications are **sensitivity** (very low limits of detection for early diagnosis), **specificity** (creating signals which are specific for the biomarker of interest), **multiplexing** (the simultaneous measurement of several biomarkers in the same sample), and **spatial resolution** (measuring biomolecular interactions on a 1–20 nm distance scale).

8.3.1 Multiplexed Diagnostics

A rapid, sensitive and specific immunoassay for biomarkers in whole blood or plasma would largely improve early diagnosis as well as therapy- and disease progression-monitoring for the benefit of clinicians (fast, easy and inexpensive) and, more importantly, of patients (uncomplicated diagnosis, better treatment). Homogeneous assays based on FRET from one dye-labeled specific anti-body (AB1) to another (AB2) within an "AB1-biomarker-AB2" immune complex are an ideal basis to meet these challenging requirements of *in vitro* diagnostics. They do not require any washing or separation steps, fast solution-phase kinetics allow short incubation times and time-resolved detection permits nearly background-free measurements. Moreover, the ratiometric format (luminescence detection of FRET donor and acceptor) offers an instantaneous suppression of sample or measurement fluctuations resulting in an extremely good reproducibility (very low coefficient of variation). Thus, homogeneous assays possess unrivalled properties for high throughput screening as well as point-of-care or small laboratory testing. The main principle of a multiplexed homogeneous FRET assay is presented in Fig. 8.2.

Due to the advantages of LTC to QD FRET (cf. Sect. 2) this FRET pair combines both very high sensitivity and multiplexed detection and is therefore theoretically very well suited for such assays. In order to practically demonstrate the performance of the LTC-QD FRET pair for diagnostic assays, we

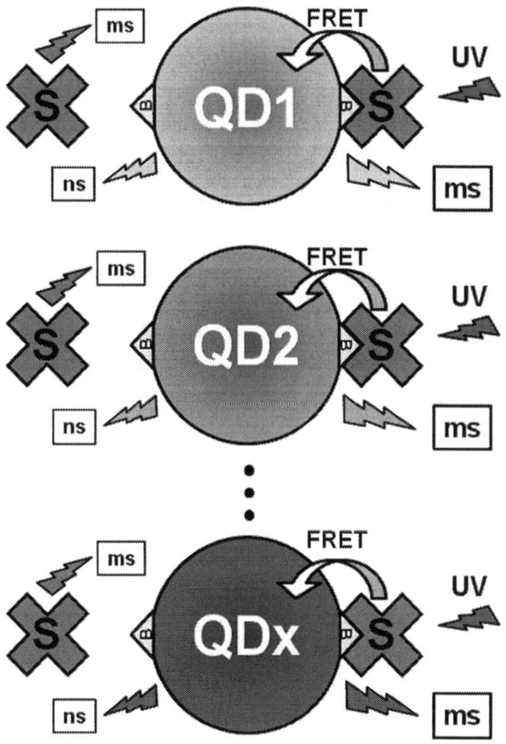

Fig. 8.3 Principle of LTC-labeled streptavidin (*S*) and biotinylated (*B*) QD multiplexed FRET assay without (*left*) and upon (*right*) binding. Copyright Wiley-VCH Verlag GmbH & Co.KGaA (Reproduced with permission from Geißler et al. 2010)

chose a bioassay, which uses molecular recognition between streptavidin (sAv) and biotin (Biot). This proof-of-concept binding system is transferable to real clinical assays using e.g. RNA, DNA, aptamer, peptide or protein based recognition. The assay principle is presented in Fig. 8.3. LTC labeled sAv binds to the different biotinylated QDs (Biot-QD) resulting in a close proximity between LTC-donors and QD-acceptors, necessary for FRET. Without binding, UV-excitation leads to long-lived (ms) LTC and short-lived (ns) QD luminescence. Upon binding, energy is transferred from LTCs to QDs resulting in luminescence quenching of LTCs and the appearance of a long-lived QD luminescence (ms). Using time-gated detection (cf. Fig. 8.1) leads to suppression of the short-lived fluorescence. The time-gated luminescence intensities at the different QD emission wavelengths (due to FRET sensitization) are proportional to the concentrations of the different FRET complexes.

In order to demonstrate the simultaneous multiplexed FRET from one LTC to five different QDs, Biot-QD was titrated to a stock solution of LTC-sAv and time-gated (50–450 μs) emission spectra were recorded (Fig. 8.4 left). The measurement of different time-gated QD emission bands is only possible due to simultaneous FRET sensitization from the long-lived LTCs. In fact, an increasing ratio of Biot-QD per LTC leads to a significant intensity increase of the time-gated QD emission spectra.

After verification of the simultaneous five-fold multiplexed FRET from LTCs to QDs the second important aspect of diagnostics, namely the sensitivity of the multiplexed assay, was investigated. For this purpose we used a modified commercial plate reader system for homogeneous FRET immunoassays (KRYPTOR™: Cezanne/BRAHMS/ThermoFisher). The KRYTOR system uses a simultaneous filter-based detection of the time-gated intensities of the LTC-donors and the QD-acceptors. The ratio of these intensities is recorded as a function of Biot-QD concentration, which leads to the characteristic FRET assay calibration curves shown in Fig. 8.4 right. Already the addition of picomolar amounts of Biot-QD to the LTC-sAv stock solution leads to a strong increase of the intensity ratio due to efficient FRET from LTCs to the five different QDs. After saturation of the Biot-sAv binding the curves level

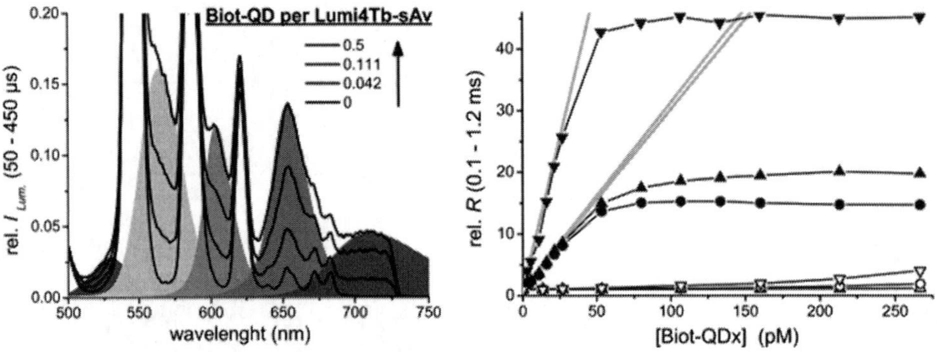

Fig. 8.4 *Left*: Time-resolved emission spectra (only the enlarged region of interest with superimposed QD steady-state spectra is shown) of increasing concentration ratios of Biot-QD per LTC-sAv (in this case the commercial LTC Lumi4Tb was used). Spectra are intensity-normalized to the 545 nm peak of the LTC. *Right*: Relative luminescence intensity ratios (as a function of Biot-QD concentration) of three different QD acceptors and the LTC donor (for better visualization only three out of five QDs are shown). *Full symbols* represent the FRET experiments (Biot-sAv binding). *Open symbols* are non-binding control experiments. Copyright Wiley-VCH Verlag GmbH & Co.KGaA (Reproduced with permission from Geißler et al. 2010)

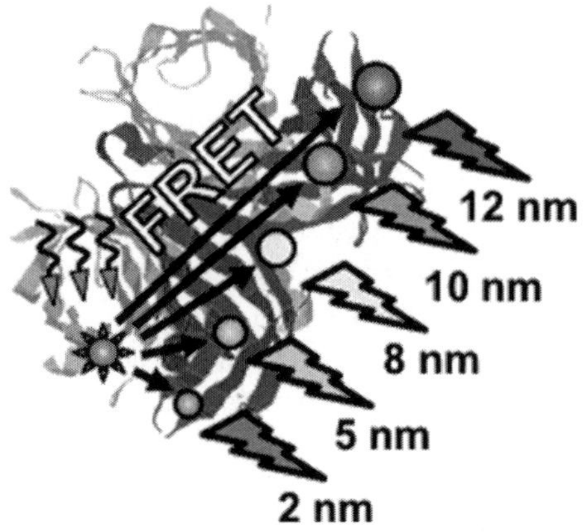

Fig. 8.5 Basic principle of a multiplexed spectroscopic ruler using a LTC-donor (labeled to one site of a protein) and several different QD-acceptors (labeled to other sites of the protein)

off. The linear parts of the calibration curves (grey lines) are used to calculate the detection limits (LODs), which are sub-picomolar (below 100 attomols) for all five different FRET systems.

In summary the LTC to QD homogeneous FRET assay offers five-fold multiplexing with sub-picomolar LODs, which is 40–240 fold better compared to the FRET assay "gold standard" (Eu-TBP-APC FRET pair) with an LOD of 24 pM for the same Biot-sAv assay.

8.3.2 Molecular Ruler/Spectroscopic Ruler

Besides the determination of concentrations in diagnostics, FRET applications also play an important role for the determination of distances in the 1–20 nm range with high accuracy far below the light diffraction limit. The "spectroscopic ruler", a name coined by Stryer and Haugland more than 40 years ago (Stryer and Haugland 1967), is frequently used for *in* and *ex vivo* studies of inter- and intramolecular interactions by spectroscopy and microscopy. Multiplexed FRET would allow the simultaneous measurement of multiple distances at a time (Fig. 8.5) for more bioanalytical information due to a possible correlation of different events.

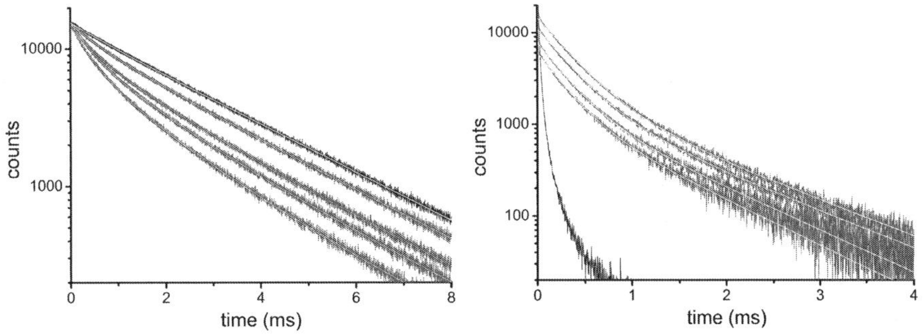

Fig. 8.6 Lifetime traces for a typical LTC-sAv-Biot-QD FRET system, and mathematical fits as thin *white lines*. *Left*: LTC luminescence decays (with increasing Biot-QD concentrations from top to bottom), showing the FRET quenching of the LTC donor. *Black curve* presents the pure (unquenched) LTC luminescence. *Right*: QD luminescence decays (with increasing Biot-QD concentrations from bottom to top), showing the FRET sensitization of the QD acceptors. *Black curve* presents the pure (not sensitized) QD luminescence at ca. 5 to 50 fold elevated QD concentration compared to the FRET measurements (*grey curves*). Copyright Wiley-VCH Verlag GmbH & Co.KGaA (Reproduced with permission from Morgner et al. 2010)

Using the same Biot-sAv binding system as explained in Fig. 8.3, we performed a time-resolved analysis of five LTC-QD FRET pairs with different emission wavelengths. This approach is on the one hand a proof-of-principle for a multiplexed spectroscopic ruler measurement (for being used in real-life systems in the future). On the other hand it can be directly used to measure the overall dimensions (shape and size) of different biocompatible QDs at sub-nanomolar concentrations under physiological conditions. Due to the large Förster distances achievable with LTC donors the FRET spectroscopic ruler can be used for measurements ranging from bare QDs up to nanocrystals with thick polymer coatings.

Multi-exponential fitting was performed for the luminescence decay curves of the different QD-acceptors as well as for the LTC-donors. As $\tau_{DA} = \tau_{AD}$ (cf. Sect. 2.1) the donor fits serve as a control of the acceptor fits. Representative curves for one of the LTC-QD FRET system are shown in Fig. 8.6. The fit results of the FRET pairs at various concentrations lead to two average decay times τ_{AD1} and τ_{AD2} with their respective pre-exponential factors. Using Eq. 8.2 (with the Förster distances calculated from spectroscopic measurements performed prior to the FRET experiments) these lifetimes can be transformed into average distances. For all five different QDs these two distances represented extremely well the two ellipsoidal axes of the nanocrystals (in agreement with the size specifications provided by Invitrogen for the applied QDs). Moreover, the pre-exponential factors could be used to determine the shape of the QDs, confirming their elongation with increasing size. The main advantages of the multiplexed spectroscopic ruler experiments compared to other analytical methods (TEM, size exclusion chromatography, etc.) are the very low concentrations and the physiological conditions. This allows a simultaneous size and shape determination of multiple QDs directly inside the biochemical sample of interest.

8.4 Conclusions and Outlook

In conclusion, LTC to QD FRET systems have many advantages for their use in bioanalysis, such as *in vitro* diagnostics and spectroscopic ruler applications. These impressive FRET tools offer optical multiplexing with low spectroscopic crosstalk, large Förster distances of more than 10 nm, multiplexed long-distance measurements and highly sensitive nearly background-free multiple parameter homogeneous immunoassays. As the presented applications of these FRET nanosensors are still in the proof-of-concept status, there is still "plenty of room at the bottom" to introduce them into challenging real-life clinical applications such as the early diagnosis of cancer or Alzheimer's disease or companion diagnostics. Moreover, these FRET pairs should be well capable of their application in cellular imaging and due to their unique photophysical properties a profound understanding of the energy transfer processes could lead to new developments beyond Förster's theory.

Acknowledgement We would like to thank the organizers of the 47th OHOLO conference for the kind invitation as well as the excellent conference in Eilat and the European Commission for financial support (FP7 Collaborative Project NANOGNOSTICS-HEALTH-F5-2009-242264).

References

Algar, W. R., & Krull, U. J. (2010). New opportunities in multiplexed optical bioanalyses using quantum dots and donor-acceptor interactions. *Analytical and Bioanalytical Chemistry, 398*, 2439–2449.

Alivisatos, A. P. (1996). Semiconductor clusters, nanocrystals, and quantum dots. *Science, 271*, 933–937.

Alivisatos, P. (2004). The use of nanocrystals in biological detection. *Nature Biotechnology, 22*, 47–52.

Boeneman, K., Mei, B. C., Dennis, A. M., Bao, G., Deschamps, J. R., Mattoussi, H., & Medintz, I. L. (2009). Sensing caspase 3 activity with quantum dot-fluorescent protein assemblies. *Journal of the American Chemical Society, 131*, 3828–3829.

Bruchez, M. P. (2005). Turning all the lights on: Quantum dots in cellular assays. *Current Opinion in Chemical Biology, 9*, 533–537.

Bruchez, M., Moronne, M., Gin, P., Weiss, S., & Alivisatos, A. P. (1998). Semiconductor nanocrystals as fluorescent biological labels. *Science, 281*, 2013–2016.

Bünzli, J.-C. G. (1989). Luminescent probes. In J.-C. G. Bünzli & G. R. Choppin (Eds.), *Lanthanide probes in life, chemical, and earth sciences: Theory and practice*. Amsterdam/New York: Elsevier.

Bünzli, J.-C. G. (2004). Luminescent lanthanide probes as diagnostic and therapeutic tools. In A. Sigel & H. Sigel (Eds.), *Metal ions in biological systems* (Vol. 42, pp. 39–75). New York: Marcel Dekker.

Bünzli, J.-C. G. (2010). Lanthanide luminescence for biomedical analyses and imaging. *Chemical Reviews, 110*, 2729–2755.

Bünzli, J.-C. G., & Piguet, C. (2005). Taking advantage of luminescent lanthanide ions. *Chemical Society Reviews, 34*, 1048–1077.

Chan, W. C. W., & Nie, S. M. (1998). Quantum dot bioconjugates for ultrasensitive nonisotopic detection. *Science, 281*, 2016–2018.

Charbonnière, L. J., & Hildebrandt, N. (2008). Lanthanide complexes and quantum dots: A bright wedding for resonance energy transfer. *European Journal of Inorganic Chemistry, 21*, 3241–3251.

Charbonnière, L. J., Hildebrandt, N., Ziessel, R. F., & Löhmannsröben, H.-G. (2006). Lanthanides to quantum dots resonance energy transfer in time-resolved FluoroImmunoAssays and luminescence microscopy. *Journal of the American Chemical Society, 128*, 12800–12809.

ChemPhysChem. (2011). Special issue on Förster resonance energy transfer. *ChemPhysChem, 12*, 421–719.

Chen, Z., Li, G., Zhang, L., Jiang, J., Li, Z., Peng, Z., & Deng, L. (2008). A new method for the detection of ATP using a quantum-dot-tagged aptamer. *Analytical and Bioanalytical Chemistry, 392*, 1185–1188.

Clapp, A. R., Medintz, I. L., Mauro, J. M., Fisher, B. R., Bawendi, M. G., & Mattoussi, H. (2004). Fluorescence resonance energy transfer between quantum dot donors and dye-labeled protein acceptors. *Journal of the American Chemical Society, 126*, 301–310.

Clapp, A. R., Medintz, I. L., Uyeda, H. T., Fisher, B. R., Goldman, E. R., Bawendi, M. G., & Mattoussi, H. (2005). Quantum dot-based multiplexed fluorescence resonance energy transfer. *Journal of the American Chemical Society, 127*, 18212–18221.

Clapp, A. R., Pons, T., Medintz, I. L., Delehanty, J. B., Melinger, J. S., Fisher, B. R., Tiefenbrunn, T., Dawson, P. E., O'Rourke, B., & Mattoussi, H. (2007). Two-photon excitation of quantum-dot-based fluorescence resonance energy transfer and its applications. *Advanced Materials, 19*, 1921–1926.

Clegg, R. M. (1996). Fluorescence resonance energy transfer. In X. F. Wang & B. Herman (Eds.), *Fluorescence imaging spectroscopy and microscopy* (Vol. 137, pp. 179–252). New York: Wiley.

Clegg, R. M. (2009). Förster resonance energy transfer – FRET. What is it, why do it, and how it's done. In T. W. J. Gadella (Ed.), *Laboratory techniques in biochemistry and molecular biology* (Vol. 33, pp. 1–57). Burlington: Academic.

Curutchet, C., Franceschetti, A., Zunger, A., & Scholes, G. D. (2008). Examining Forster energy transfer for semiconductor nanocrystalline quantum dot donors and acceptors. *Journal of Physical Chemistry C, 112*, 13336–13341.

Dale, R. E., & Eisinger, J. (1974). Intramolecular distances determined by energy-transfer-dependence on orientational freedom of donor and acceptor. *Biopolymers, 13*, 1573–1605.

Dale, R. E., Eisinger, J., & Blumberg, W. E. (1979). Orientational freedom of molecular probes – Orientation factor in intra-molecular energy-transfer. *Biophysical Journal, 26*, 161–193.

Diamandis, E. P. (1988). Immunoassays with time-resolved fluorescence spectroscopy – Principles and applications. *Clinical Biochemistry, 21*, 139–150.

Didenko, V. V. (2001). DNA probes using Fluorescence Resonance Energy Transfer (FRET): Designs and applications. *Biotechniques, 31*, 1106–1116.

Fairclough, R. H., & Cantor, C. R. (1978). The use of singlet-singlet energy transfer to study macromolecular assemblies. In C. H. W. Hirs & S. N. Timasheff (Eds.), *Methods in enzymology* (48, pp. 347–379). New York: Academic.

Förster, T. H. (1948). Zwischenmolekulare Energiewanderung und Fluoreszenz. *Annalen der Physik, 2*, 55–751.

Förster, T. H. (1949). Experimentelle und theoretische Untersuchung des zwischenmolekularen Übergangs von Elektronenanregungsenergie. *Zeitschrift für Naturforschung, 4*, 321–327.

Förster, T. H. (1959). 10th Spiers Memorial Lecture – Transfer mechanisms of electronic excitation. *Discussions of the Faraday Society, 27*, 7–17.

Förster, Th. (1965). Delocalized excitation and excitation transfer. In O. Sinanoglu (Ed.), *Modern quantum chemistry. Istanbul Lectures. Part III* (pp. 93–137). New York/London: Academic.

Fung, B. K. K., & Stryer, L. (1978). Surface density determination in membranes by fluorescence energy transfer. *Biochemistry, 17*, 5241–5248.

Geißler, D., & Hildebrandt, N. (2011). Lanthanide complexes in FRET applications. *Current Inorganic Chemistry, 1*, 17–35.

Geißler, D., Charbonnière, L. J., Ziessel, R. F., Butlin, N. G., Löhmannsröben, H.-G., & Hildebrandt, N. (2010). Quantum dot biosensors for ultra-sensitive multiplexed diagnostics. *Angewandte Chemie, International Edition, 49*, 1396–1401.

Gill, R., Zayats, M., & Willner, I. (2008). Semiconductor quantum dots for bioanalysis. *Angewandte Chemie, International Edition, 47*, 7602–7625.

Gudgin Dickson, E. F., Pollak, A., & Diamandis, E. P. (1995). Ultrasensitive bioanalytical assays using time-resolved fluorescence detection. *Pharmacology and Therapeutics, 66*, 207–235.

Haas, E., Katchalskikatzir, E., & Steinberg, I. Z. (1978). Effect of the orientation of donor and acceptor on the probability of energy transfer involving electronic transitions of mixed polarization. *Biochemistry, 17*, 5064–5070.

Härmä, H., Soukka, T., Shavel, A., Gaponik, N., & Weller, H. (2007). Luminescent energy transfer between cadmium telluride nanoparticle and lanthanide(III) chelate in competitive bioaffinity assays of biotin and estradiol. *Analytica Chimica Acta, 604*, 177–183.

Hemmilä, I., & Laitala, V. (2005). Progress in lanthanides as luminescent probes. *Journal of Fluorescence, 15*, 529–542.

Hemmilä, I., & Mukkala, V.-M. (2001). Time-resolution in fluorometry technologies, labels, and applications in bioanalytical assays. *Critical Reviews in Clinical Laboratory Sciences, 38*, 441–519.

Hildebrandt, N., & Löhmannsröben, H.-G. (2007). Quantum dot nanocrystals and supramolecular lanthanide complexes – energy transfer systems for sensitive in vitro diagnostics and high throughput screening in chemical biology. *Current Chemical Biology, 1*, 167–186.

Hildebrandt, N., Charbonnière, L. J., Beck, M., Ziessel, R. F., & Löhmannsröben, H.-G. (2005). Quantum dots as efficient energy acceptors in a time-resolved fluoroimmunoassay. *Angewandte Chemie, International Edition, 44*, 7612–7615.

Hildebrandt, N., Charbonnière, L. J., & Löhmannsröben, H. -G. (2007) Time-resolved analysis of a highly sensitive Förster Resonance Energy Transfer (FRET) immunoassay using terbium complexes as donors and quantum dots as acceptors. *Journal of Biomedicine and Biotechnology*, Article ID 79169: 6 p.

Hofmann, H., Hillger, F., Pfeil, S. H., Hoffmann, A., Streich, D., Haenni, D., Nettels, D., Lipman, E. A., & Schuler, B. (2010). Single-molecule spectroscopy of protein folding in a chaperonin cage. *Proceedings of the National Academy of Sciences of the United States of America, 107*, 11793–11798.

Jamieson, R., Bakhshi, D., Petrova, R., Pocock, M. I., & Seifalian, A. M. (2007). Biological applications of quantum dots. *Biomaterials, 28*, 4717–4732.

Jares-Erijman, E. A., & Jovin, T. M. (2003). FRET imaging. *Nature Biotechnology, 21*, 1387–1395.

Jares-Erijman, E. A., & Jovin, T. M. (2009). Reflections on FRET imaging: Formalism, probes and implementation. In T. W. J. Gadella (Ed.), *Laboratory techniques in biochemistry and molecular biology* (33, pp. 475–517). Amsterdam: Elsevier.

Katz, E., & Willner, I. (2004). Integrated nanoparticle-biomolecule hybrid systems: Synthesis, properties and applications. *Angewandte Chemie, International Edition, 43*, 6042–6108.

Lakowicz, J. R. (1999). *Principles of fluorescence spectroscopy* (2nd ed.). New York: Kluwer Academic/Plenum.

Lakowicz, J. R., & Geddes, C. D. (1991). *Topics in fluorescence spectroscopy*. New York: Plenum Press.
Medintz, I. L., Clapp, A. R., Mattoussi, H., Goldman, E. R., Fisher, B., & Mauro, J. M. (2003). Self-assembled nanoscale biosensors based on quantum dot FRET donors. *Nature Materials, 2*, 630–638.
Medintz, I. L., Uyeda, H. T., Goldman, E. R., & Mattoussi, H. (2005). Quantum dot bioconjugates for imaging, labelling and sensing. *Nature Materials, 4*, 435–446.
Medintz, I. L., Mattoussi, H., & Clapp, A. R. (2008). Potential clinical applications of quantum dots. *International Journal of Nanomedicine, 3*, 151–167.
Michalet, X., Pinaud, F. F., Bentolila, L. A., Tsay, J. M., Doose, S., Li, J. J., Sundaresan, G., Wu, A. M., Gambhir, S. S., & Weiss, S. (2005). Quantum dots for live cells, in vivo imaging, and diagnostics. *Science, 307*, 538–544.
Morgner, F., Geißler, D., Stufler, S., Butlin, N. G., Löhmannsröben, H.-G., & Hildebrandt, N. (2010). A quantum dot-based molecular ruler for multiplexed optical analysis. *Angewandte Chemie, International Edition, 49*, 7570–7574.
Morrison, L. E. (1988). Time-resolved detection of energy-transfer – Theory and application to immunoassays. *Analytical Biochemistry, 174*, 101–120.
Murphy, C. J., & Coffer, J. L. (2002). Quantum dots: A primer. *Applied Spectroscopy, 56*, 16A–27A.
Penn, S. G., He, L., & Natan, M. J. (2003). Nanoparticles for bioanalysis. *Current Opinion in Chemical Biology, 7*, 609–615.
Raymo, F. M., & Yildiz, I. (2007). Luminescent chemosensors based on semiconductor quantum dots. *Physical Chemistry Chemical Physics, 9*, 2036–2043.
Reviews in Molecular Biotechnology. (2002). Special issue on FRET. *Reviews in Molecular Biotechnology, 82*, 177–300.
Richardson, F. S. (1982). Terbium(III) and Europium(III) Ions as luminescent probes and stains for biomolecular systems. *Chemical Reviews, 82*, 541–552.
Roberti, M. J., Giordano, L., Jovin, T. M., & Jares-Erijman, E. A. (2011). FRET imaging by k_t/k_f. *ChemPhysChem, 123*, 563–566.
Sammes, P. G., & Yahioglu, G. (1996). Modern bioassays using metal chelates as luminescent probes. *Natural Product Reports, 13*, 1–28.
Sapsford, K. E., Berti, L., & Medintz, I. L. (2006). Materials for fluorescence resonance energy transfer analysis: Beyond traditional donor-acceptor combinations. *Angewandte Chemie, International Edition, 45*, 4562–4588.
Schuler, B., Lipman, E. A., Steinbach, P. J., Kumke, M., & Eaton, W. A. (2005). Polyproline and the "Spectroscopic Ruler" revisited with single-molecule fluorescence. *Proceedings of the National Academy of Sciences of the United States of America, 102*, 2754–2759.
Selvin, P. R. (2000). The renaissance of fluorescence resonance energy transfer. *Nature Structural Biology, 7*, 730–734.
Selvin, P. R. (2002). Principles and biophysical applications of lanthanide-based probes. *Annual Review of Biophysics and Biomolecular Structure, 31*, 275–302.
Somers, R. C., Bawendi, M. G., & Nocera, D. G. (2007). CdSe nanocrystal based chem-/bio- sensors. *Chemical Society Reviews, 36*, 579–591.
Steinberg, I. Z. (1971). Long-range nonradiative transfer of electronic excitation energy in proteins and polypeptides. *Annual Review of Biochemistry, 40*, 83–114.
Stryer, L. (1978). Fluorescence energy-transfer as a spectroscopic ruler. *Annual Review of Biochemistry, 47*, 819–846.
Stryer, L., & Haugland, R. P. (1967). Energy transfer: A spectroscopic ruler. *Proceedings of the National Academy of Sciences of the United States of America, 58*, 716–726.
Szollosi, J., Damjanovich, S., & Matyus, L. (1998). Application of fluorescence resonance energy transfer in the clinical laboratory: Routine and research. *Cytometry, 34*, 159–179.
Van der Meer, B. W., Coker, G., & Simon Chen, S. Y. (1994). *Resonance energy transfer: Theory and data*. New York: Cambridge, VCH.
Wu, P. G., & Brand, L. (1994). Resonance energy-transfer - methods and applications. *Analytical Biochemistry, 218*, 1–13.
Yin, Y., & Alivisatos, A. P. (2005). Colloidal nanocrystal synthesis and the organic-inorganic interface. *Nature, 437*, 664–670.
Yuan, J., & Wang, G. (2005). Lanthanide complex-based fluorescence label for time-resolved fluorescence bioassay. *Journal of Fluorescence, 15*, 559–568.

Chapter 9
Assembly and Microscopic Characterization of DNA Origami Structures

Max Scheible, Ralf Jungmann, and Friedrich C. Simmel

Abstract DNA origami is a revolutionary method for the assembly of molecular nanostructures from DNA with precisely defined dimensions and with an unprecedented yield. This can be utilized to arrange nanoscale components such as proteins or nanoparticles into pre-defined patterns. For applications it will now be of interest to arrange such components into functional complexes and study their geometry-dependent interactions. While commonly DNA nanostructures are characterized by atomic force microscopy or electron microscopy, these techniques often lack the time-resolution to study dynamic processes. It is therefore of considerable interest to also apply fluorescence microscopic techniques to DNA nanostructures. Of particular importance here is the utilization of novel super-resolved microscopy methods that enable imaging beyond the classical diffraction limit.

Keywords DNA nanotechnology • DNA origami • AFM • Super-resolution imaging

9.1 Introduction

One of the main inspirations for nanotechnology comes from the remarkable ability of biological systems to self-assemble – to form structure without external guidance, merely directed by local molecular information. Among the most fascinating and paradigmatic examples in this respect is the molecular recognition between two strands of DNA. DNA molecules are oligomers or polymers of subunits called nucleotides that carry the "bases" adenine (A), guanine (G), cytosine (C) and thymine (T). These bases can occur in any order within a DNA molecule, and distinct sequences of bases correspond to distinct DNA molecules. Importantly, bases can bind to each other via hydrogen bonds. In the famous Watson-Crick bonding scheme, adenine binds to thymine and guanine binds to cytosine. When two DNA molecules have Watson-Crick complementary base sequences, they can line up and bind to each other to form a double-helical duplex molecule that is held together by base-pairing and so-called "stacking" interactions. In biology, the base sequences of DNA molecules contain information that is processed by the molecular genetic machinery. But DNA also is an interesting biopolymer in its own right. For instance, DNA is a highly charged molecule surrounded by a structured cloud of

M. Scheible (✉) • R. Jungmann • F.C. Simmel
Lehrstuhl für Bioelektronik, Physics Department & ZNN/WSI, Technische Universität München, Am Coulombwall 4a, Garching D-85748, Germany

E. Zahavy et al. (eds.), *Nano-Biotechnology for Biomedical and Diagnostic Research*, Advances in Experimental Medicine and Biology 733, DOI 10.1007/978-94-007-2555-3_9,

counter-ions. On a short length scale, DNA duplexes are relatively rigid, rod-like molecules, while on a long scale their "random" polymer nature becomes noticeable. In the past decades these and many other remarkable properties of DNA as an "information-carrying" polymer have been utilized to generate non-biological increasingly complicated molecular structures and devices. In the following paragraphs, we give a short introduction into the field with a particular emphasis on biophysical characterization tools for the study of DNA-based nanostructures.

9.2 DNA as a Material for Molecular Self-Assembly

9.2.1 DNA Self-Assembly

Already in the early 1980s, Nadrian Seeman had the vision that DNA molecules should be used outside of their biological context, namely as building blocks for molecular "self-assembly". This could be used, e.g., to periodically arrange proteins that would be hard to crystallize otherwise in order to aid structure determination by X-ray crystallography (Seeman 1982). After the realization of molecular objects such as a cube (Chen and Seeman 1991) or a truncated octahedron (Zhang and Seeman 1994) made from DNA, two-dimensional molecular lattices were realized that were essentially based on the "four-way" or "Holliday" junction motif (Winfree et al. 1998). In biological Holliday "intermediates", two homologous double-stranded DNA molecules (carrying essentially the same sequence) exchange one of their strands, forming a structure with four "arms" that are connected at the branch point. Biologically, this occurs in DNA repair or recombination processes. It turns out that four-way junctions are quite flexible, and therefore are not very useful for the assembly of ordered structures per se. However, Seeman and coworkers invented structures such as the double or triple crossover (DX and TX), in which several four-way junctions are connected together to form rigid building blocks, often termed "tiles". Multiple crossover structures form the basic unit of most of the DNA structures realized until then (a notable exception is a DNA motif designed by et al. (Hamada and Murata 2009)), and they are also found in the origami structures discussed below.

Recently, Seeman's original vision to use DNA nanostructures as a tool for structural biology actually seems to become reality: Seeman and coworkers demonstrated the assembly of a self-assembled three-dimensional DNA crystal with dimensions of several 100 μm (Zheng et al. 2009). William Shih and collaborators demonstrated that DNA origami-based liquid crystals can be used to support nuclear magnetic resonance (NMR) experiments on proteins (Douglas et al. 2007). Finally, Turberfield and coworkers showed that arrangement of proteins on a 2D DNA lattice can strongly improve structure determination by cryo-electron microscopy (Selmi et al. 2011).

9.2.2 DNA Origami: Concept and Design of Structures

In 2006 Paul Rothemund experimentally demonstrated a revolutionary DNA-based assembly method (Rothemund 2006). It differs from previous tile assembly techniques in one fundamental aspect. Rather than assembling a large structure from many short molecules, one very large molecule (the "scaffold strand") is mixed with a multitude of short "staple strands". These staple strands connect different sections of the scaffold strand by hybridization, forming a network of strand crossover structures that forces the scaffold to adopt a defined shape. In origami structures, the scaffold strand is "routed" through the desired shape, and this can be specified by the choice of appropriate staple sequences. It turns out that this assembly scheme is highly efficient (large yield of correctly formed structures)

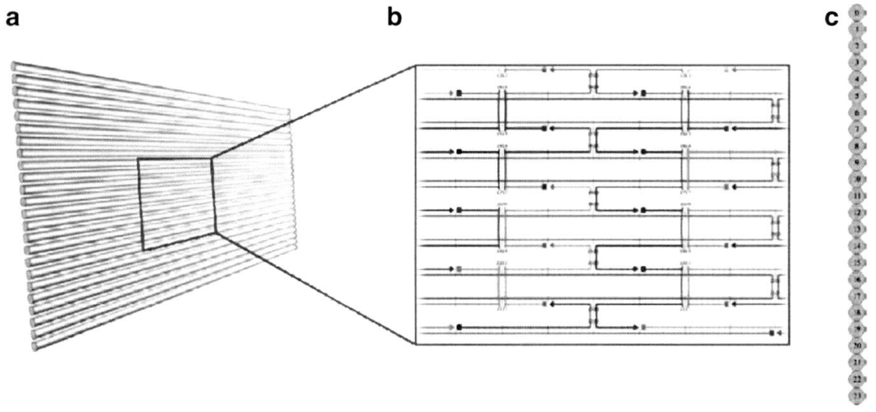

Fig. 9.1 (**a**) Illustration of a rectangular DNA origami structure designed with caDNAno (freely available for download at http://cadnano.org) (Douglas et al. 2009b). It consists of 24 antiparallel double helices with a total size of $100 \times 70 \times 2$ nm³. (**b**) Detailed section of the design showing the scaffold strand in *light blue* and colored staple strands connecting up to three double helices by interhelical junctions. (**c**) Side view of the 24 double helices in planar configuration

and robust (folding occurs even when some strands are missing). In contrast to tile-based assembly, folding of the long strand occurs "intramolecularly" and hence seems to be kinetically favorable. Furthermore, it is not necessary to obey any stoichiometric constraints – staple strands are actually added in considerable excess over the scaffold strands, which leads to faster hybridization and also enables "healing" of the structures – incorrectly bound staples presumably are displaced by better matching strands, maximizing the total number of base-pairs formed. One potential drawback (as with most other DNA assembly techniques) is the need for thermal annealing – typically, staple/scaffold mixtures have to be heated up to high temperatures and slowly cooled down over the course of hours (for 2D structures) or even days (for 3D origami). Thermal annealing can be avoided, e.g., by the addition and slow removal of denaturing agents such as formamide (Jungmann et al. 2008; Högberg et al. 2009). For future applications, e.g., for the assembly of origami-like structures within cells, isothermal – possibly enzymatically supported – processes would be desirable.

With the highly predictable and efficient assembly of DNA origami structures, DNA assembly has taken a major step towards "biomolecular engineering". In fact, several groups have already developed software packages that support the design of origami structures with a graphical user interface similar to computer-aided design (CAD) systems for engineers (e.g., caDNAno (Douglas et al. 2009b) and SARSE (Andersen et al. 2008)). This is a quite remarkable and unprecedented development for self-assembly and molecular nanotechnology! One example for an origami design made with caDNAno is shown in Fig. 9.1. Here, the program was used to choose the staples sequences needed to fold the origami scaffold into a long rectangular DNA structure. The DNA strands faithfully created the target structure with a very high yield, as evidenced by AFM imaging. For a didactic step-by-step introduction into the design and assembly of origami structures, the reader is referred to Refs. (Rothemund 2006; Castro et al. 2011).

9.3 Modification of Origami Structures

One of the major issues for the origami field in the near future will be the demonstration of "real" applications. Researchers envision origami-based drugs and delivery systems, the arrangement of nanoparticles for applications in nanoelectronics and nanoplasmonics, the realization of artificial enzyme cascades and complexes, and even molecular assembly lines (Gu et al. 2010). In all cases, efficient methods for the attachment of nanoscale components on origami structures are required. There are a variety of obvious strategies, e.g. the direct incorporation of DNA conjugates or hybridization of DNA conjugates to staple extensions on the origami. Direct incorporation during folding of the origami structures works well for small and thermally stable molecules, e.g. for labeling of origami structures with fluorescent dyes. Subsequent hybridization to staple extensions has been shown to work well for the detection of RNA molecules on an origami sheet, or for the attachment of oligo-coated metal nanoparticles (Pal et al. 2010; Hung et al. 2010). The main advantage of such hybridization-based methods is the sequence-specificity of attachment. In principle, a large variety of different components can be labeled with distinct DNA sequences and directed to unique binding sites on the origami. In practice, direct hybridization often suffers from low yield or is not applicable simply because oligo-labeled compounds are not available or feasible.

Other attachment strategies utilize molecular recognition by specific binders such as aptamers or antibodies. DNA aptamers – if available – have the particular advantage that they can be directly incorporated into origami structures as extended staple strands (Ke et al. 2008).

One of the most often used strategies, however, is the utilization of biotinylated staple strands for the attachment of streptavidin or conjugates thereof. This has been demonstrated for arrangement of streptavidin proteins on origami into patterns (Kuzyk et al. 2009), for the binding of streptavidin-coated quantum dots, or streptavidin-enzyme conjugates (cf. Fig. 9.2).

Even though the streptavidin-biotin system with its extremely high affinity (K_d =1 fM) typically works extremely well, it is not sufficient for the attachment of multiple distinct components on a single origami structure. Therefore, a variety of research groups currently explore alternative – "orthogonal" – binding methods. For instance, Niemeyer and coworkers recently demonstrated the utilization of self-ligating protein tags such as the Snap-tag and the HaloTag for this purpose (Sacca et al. 2010). Finally, another possibility for immobilization of components is the utilization of enzymatic methods. For instance, terminal transferase can be used to enzymatically label staple strands with modified ddNTPs (Jahn et al. 2011). A variety of examples for site-specific immobilization of nano-components utilizing the streptavidin/biotin system on origami rectangles are shown in Fig. 9.2.

9.4 Characterization of DNA Origami Structures

9.4.1 AFM Imaging

One of the most commonly used techniques to image DNA origami is atomic force microscopy (AFM). In AFM, a microscopic silicon or silicon nitride beam with a sharp tip (the "cantilever") is brought into contact with a surface. The resulting deflection of the cantilever can be measured and correlated to the height (or other physical properties) of surface structures. By scanning the tip over the surface, images of the surface can be obtained that represent a map of the physical property under study. For the characterization of biological molecules, the AFM is most frequently operated in the intermittent contact (or tapping™) mode, in which an oscillating tip is scanned over a surface. The distance of the cantilever is chosen such that the tip only briefly touches the surface, resulting in relatively "gentle" imaging. The tapping mode also works in liquid and is therefore well suitable for imaging of biomolecules in biological buffers.

As native dsDNA has a diameter of only 2 nm, AFM imaging of DNA structures requires an extremely flat substrate. Most often, atomically flat sheets of mica are utilized to which DNA molecules stick electrostatically in the presence of divalent cations. As examples, in Figs. 9.2 and 9.3, DNA origami structures are visualized in AFM height contrast images. A variety of other measurement

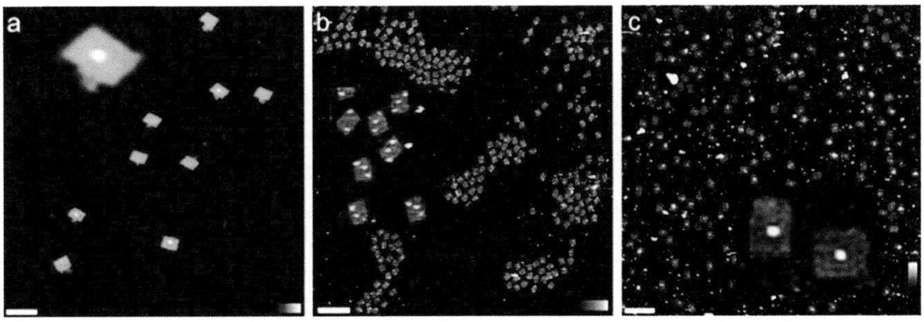

Fig. 9.2 Demonstration of various labeling approaches by the biotin-streptavidin-system. (**a**) AFM height contrast image of a quantum dot coated with streptavidin bound to a biotinylated staple strand in the middle of an origami structure, length scale: 200 nm, height scale: 10 nm. (**b**) DNA origami with three streptavidin labels arranged in a diagonal configuration on the origami structure, length scale: 400 nm, height scale: 16 nm. (**c**) Origami structures modified with horseradish peroxidase conjugated streptavidin via biotinylated staple strands in the middle of the structure demonstrate the feasibility of DNA origami-enzyme complexes, length scale: 300 nm, height scale: 10 nm

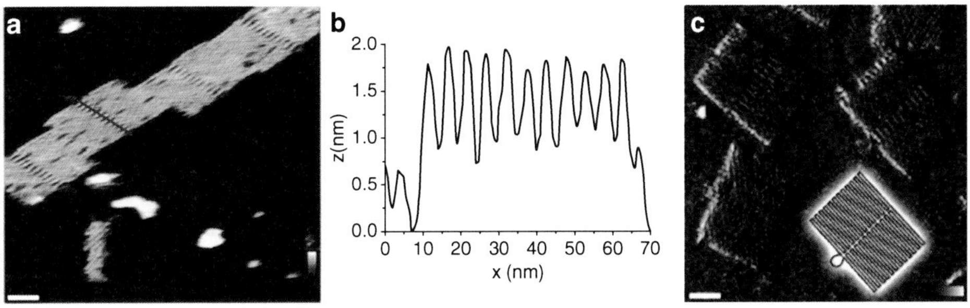

Fig. 9.3 (**a**) AFM height contrast image of 4 DNA origami units highlighting sub-structural details like the bridged seam in the middle of the structure, length scale: 30 nm, height scale 6 nm. (**b**) A cross section profile of the bridged seam (marked *red* in (**a**)) demonstrates the high resolution potential by AFM imaging. (**c**) AFM amplitude error image of DNA origami structures, length scale: 30 nm, height scale: 150 mV (**a**, **b** reproduced from Jungmann et al. 2008, © American Chemical Society)

modes are available that contain further information about the sample. E.g., so-called phase contrast images contain information about the mechanical properties of the sample. With AFM height differences well below 1 nm can be detected, and – depending on tip size and quality – even sub-structural features of origami assemblies can be visualized. For instance, in Fig. 9.3a rectangular origami structures are displayed that contain a seam-like structure in the middle (indicated by the red line). This structure results from the particular design of this rectangular origami structure. Each line of the seam actually represents two double helices connecting the two halves of the rectangle. A cross-section profile through the seam structure is shown in Fig. 9.3b.

As already mentioned in Sect. 9.3, one of the most exciting features of DNA origami is its potential for the spatial arrangement of nanoscale components in a precise, sequence-addressable manner. The sensitive height contrast of AFM allows imaging of such modifications. As an example, Fig. 9.3c shows origami structures modified with DNA dumbbell structures that are visible as elevated features protruding from the origami sheets.

An extremely interesting development of recent years is the development of video-rate AFM that also allows imaging of dynamic processes. This has been recently demonstrated by imaging the action of enzyme on substrates immobilized on origami structures (Endo et al. 2010).

9.4.2 Electron Microscopy

AFM is very well suited for studying biological assemblies in biological buffers and offers a good height resolution of surface features. However, the lateral resolution under such conditions is typically an order of magnitude worse (i.e., a few nm). This becomes a problem for three-dimensional DNA assemblies that are typically more compact and less extended in the lateral dimensions than 2D structures. For this reason, the method of choice for characterization of 3D origami structures has become transmission electron microscopy (TEM). In TEM, an energetic electron beam is used to illuminate the sample. The small de Broglie wavelength of the electrons enables much higher resolutions compared to optical microscopy. Even at 100 kV acceleration voltages (which are typically used for biological samples), a lateral resolution less than 1 nm is easily achieved. The power of TEM has been recently impressively demonstrated by imaging of the substructure of a variety of 3D origami structures (Douglas et al. 2009a; Dietz et al. 2009; Liedl et al. 2010), and the structural analysis of even smaller DNA objects not based on origami (He et al. 2008; Kato et al. 2009). The main disadvantage of electron microscopy compared to AFM, however, is the tedious and invasive sample preparation process. Typically, chemical or cryo-fixation as well as staining protocols (with uranyl salts) have to be used in advance, while imaging takes place under vacuum conditions.

9.4.3 Fluorescence Microscopy of DNA Nanostructures

Static and invasive imaging techniques such as atomic force microscopy or electron microscopy usually are not capable of imaging fast dynamical processes. Therefore light and fluorescence microscopy are prevalent in the biological sciences, as it is possible to observe the objects under study in real-time and under native conditions.

For fluorescence microscopy the target objects have to be modified with fluorescent dyes – something that is readily achieved with DNA nanostructures, as a large variety of fluorescent labels for DNA is available commercially. As light microscopy in general, however, fluorescence microscopy is constrained by the "diffraction limit", which states that two light emitting points cannot be distinguished from each other when they have a distance of the order of the wavelength of light or smaller. For this reason, fluorescence microscopy has been only seldomly used for the characterization of DNA nanostructures, as in most cases the structures were simply too small. Only assemblies with sufficient size (i.e., on the order of several micrometers or more) are amenable to standard fluorescence microscopic analysis. Some notable examples are the characterization of the assembly dynamics of fluorescently labeled DNA nanotubes (Ekani-Nkodo et al. 2004), melting studies of DNA nanotubes (Sobey et al. 2009), and the imaging of highly extended DNA lattices (He et al. 2005).

Another issue in fluorescence microscopy is the high signal noise due to background fluorescence. In order to be able to detect single molecules as well as to improve the signal to noise ratio, typically total internal reflection fluorescence microscopy (TIRFM) is used (Axelrod et al. 1984). For TIRFM a laser beam is used to irradiate the boundary region between cover glass and sample solution at an angle beyond the critical angle for total internal reflection. At the reflection point, an evanescent field is generated, which decays exponentially into the sample solution. Hence only fluorophores located within 100–200 nm in the direct proximity of the glass surface are excited, efficiently reducing background fluorescence. The ability to monitor the fluorescence of single molecules is an important ingredient of a variety of novel microscopy methods that actually allow for optical imaging beyond the diffraction limit, a development that will have important consequences also for studies on DNA-based nanostructures.

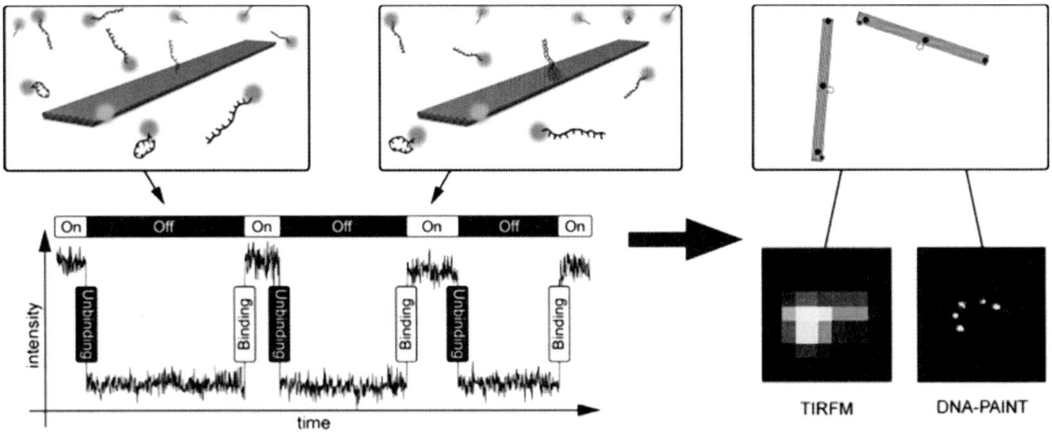

Fig. 9.4 The DNA-PAINT principle: a freely diffusing fluorescently labeled oligomer (*red*) can bind transiently to its complementary counterpart, which is a "docking strand" on the origami structure. Control over the hybridization kinetics enables the detection of single bright state events of one transiently binding fluorophore as shown in the intensity vs. time trace. E.g., DNA origami structures with three docking strands can be analyzed by stochastic signal read-out resolving single strand positions beyond the diffraction limit (Reproduced from Jungmann et al. 2010, © American Chemical Society)

9.4.4 Super-Resolution Microscopy of DNA Origami

In recent years, several techniques were developed to circumvent the diffraction limit of far-field optical microscopy and to obtain a resolution below the diffraction limit, namely super-resolution. One important requirement for super-resolution is the ability to localize single fluorophores. This can either be fulfilled by targeted or stochastic readout techniques (Hell 2009; Vogelsang et al. 2010), both providing a resolution of up to 20 nm in three dimensions (Schermelleh et al. 2010).

We here focus on techniques that use readout schemes based on the stochastic analysis of bright and dark states of switchable fluorophores as, e.g., in photo-activated localization microscopy (Betzig et al. 2006) or stochastic optical reconstruction microscopy (Rust et al. 2006; Huang et al. 2009). While PALM and STORM rely on photoactivatable proteins or single-molecule switches (e.g. Cy3 and Cy5 molecules in spatial proximity) to stochastically switch fluorophores between bright and dark states, blink microscopy uses a special buffer system to induce the switching. This system consists of a mixture of a reducing and oxidizing agent (Vogelsang et al. 2008) and is able to specifically control fluorescence on- and off-states of single fluorophores.

Blink microscopy was successfully combined with the unique nanoscale addressability of DNA origami to build a nanoscopic ruler for super-resolution microscopy (Steinhauer et al. 2009). This was the first application of a super-resolution technique for imaging DNA-based nanostructures such as DNA origami.

Apart from imaging using microscopy, fluorescence can also be used to probe interactions on the single-molecule level. For applications of DNA-based nanostructures as functional biomaterials, the study of reaction kinetics and dynamic processes is extremely important. A single-molecule assay related to the PAINT (Sharonov and Hochstrasser 2006) method allows the study of binding and unbinding kinetics on DNA origami (Jungmann et al. 2010). In DNA-PAINT, a DNA origami structure is modified with a single stranded staple extension (docking strand) and is immobilized on a glass surface (cf. Fig. 9.4). A fluorescently labeled "imager" strand, which is complementary to the docking strand on the DNA origami, diffuses freely in solution and is able to transiently bind to the origami structure. A total internal reflection microscope is used to detect these single binding and unbinding events and from the extracted times for the bound and unbound state the reaction kinetics can be obtained.

The unique tunability of DNA hybridization interactions was used to control the bound and unbound state. Binding and unbinding can be interpreted as a virtual blinking of fluorescent molecules as in STORM or blink microscopy. DNA-PAINT can therefore be readily used as a super-resolution technique. By applying a 2-D Gaussian fit to the point spread function of each binding event, the precise localization of the docking strand positions on DNA origami structures could be resolved up to a resolution of about 30 nm (cf. Fig. 9.4).

With an appropriate labeling protocol, DNA-PAINT could be also applied to bioimaging, e.g., by conjugation of DNA to an antibody or by using DNA aptamers that bind to specific biomolecules. Further optimization of the self-quenching of the probe in its unbound state or even triggered self-assembly of imaging probes could accelerate super-resolution imaging and facilitate measurements with a simple epifluorescence setup.

9.5 Conclusion and Outlook

DNA has proven to be a versatile material for the assembly of intricate molecular structures, which may be used to precisely arrange nanoscale components such as proteins or nanoparticles in two and even three dimensions. In particular the novel origami technique enables the assembly of well-defined molecular objects with an unprecedented yield. The origami assembly process is so reliable that it lends itself to automation – computer-aided design and automated assembly have already been shown feasible. With such a robust technique at hand, it will now be important to find good applications for DNA-assembled nanostructures, which will involve characterization of the nanocomponents on the origami structures, their interactions, dynamics and motion. This will require the application of sophisticated characterization tools. In addition to static methods such as electron microscopy and atomic force microscopy, novel techniques such as fast-scanning AFM and super-resolution fluorescence microscopy are expected to play an important role in this context in the near future.

References

Andersen, E. S., Dong, M. D., Nielsen, M. M., Jahn, K., Lind-Thomsen, A., Mamdouh, W., Gothelf, K. V., Besenbacher, F., & Kjems, J. (2008). DNA origami design of dolphin-shaped structures with flexible tails. *ACS Nano, 2*, 1213–1218.

Axelrod, D., Burghardt, T. P., & Thompson, N. L. (1984). Total internal-reflection fluorescence. *Annual Review of Biophysics and Bioengineering, 13*, 247–268.

Betzig, E., Patterson, G. H., Sougrat, R., Lindwasser, O. W., Olenych, S., Bonifacino, J. S., Davidson, M. W., Lippincott-Schwartz, J., & Hess, H. F. (2006). Imaging intracellular fluorescent proteins at nanometer resolution. *Science, 313*, 1642–1645.

Castro, C. E., Kilchherr, F., Kim, D.-N., Shiao, E. L., Wauer, T., Wortmann, P., Bathe, M., & Dietz, H. (2011). A primer to scaffolded DNA origami. *Nature Methods, 8*, 221–229.

Chen, J. H., & Seeman, N. C. (1991). Synthesis from DNA of a molecule with the connectivity of a cube. *Nature, 350*, 631–633.

Dietz, H., Douglas, S. M., & Shih, W. M. (2009). Folding DNA into twisted and curved nanoscale shapes. *Science, 325*, 725–730.

Douglas, S. M., Chou, J. J., & Shih, W. M. (2007). DNA-nanotube-induced alignment of membrane proteins for NMR structure determination. *Proceedings of the National Academy of Sciences of the United States of America, 104*, 6644–6648.

Douglas, S. M., Dietz, H., Liedl, T., Högberg, B., Graf, F., & Shih, W. M. (2009a). Self-assembly of DNA into nanoscale three-dimensional shapes. *Nature, 459*, 414–418.

Douglas, S. M., Marblestone, A. H., Teerapittayanon, S., Vazquez, A., Church, G. M., & Shih, W. M. (2009b). Rapid prototyping of 3D DNA-origami shapes with caDNAno. *Nucleic Acids Research, 37*, 5001–5006.

Ekani-Nkodo, A., Kumar, A., & Fygenson, D. K. (2004). Joining and scission in the self-assembly of nanotubes from DNA tiles. *Physical Review Letters, 93*, 268301.

Endo, M., Katsuda, Y., Hidaka, K., & Sugiyama, H. (2010). Regulation of DNA methylation using different tensions of double strands constructed in a defined DNA nanostructure. *Journal of the American Chemical Society, 132*, 1592–1597.

Gu, H., Chao, J., Xiao, S.-J., & Seeman, N. C. (2010). A proximity-based programmable DNA nanoscale assembly line. *Nature, 465*, 202–205.

Hamada, S., & Murata, S. (2009). Substrate-assisted assembly of interconnected single-duplex DNA nanostructures13. *Angewandte Chemie, International Edition, 48*, 6820–6823.

He, Y., Tian, Y., Chen, Y., Deng, Z. X., Ribbe, A. E., & Mao, C. D. (2005). Sequence symmetry as a tool for designing DNA nanostructures. *Angewandte Chemie, International Edition, 44*, 6694.

He, Y., Ye, T., Su, M., Zhang, C., Ribbe, A. E., Jiang, W., & Mao, C. D. (2008). Hierarchical self-assembly of DNA into symmetric supramolecular polyhedra. *Nature, 452*, 198–202.

Hell, S. W. (2009). Microscopy and its focal switch. *Nature Methods, 6*, 24–32.

Högberg, B., Liedl, T., & Shih, W. M. (2009). Folding DNA origami from a double-stranded source of scaffold. *Journal of the American Chemical Society, 131*, 9154–9155.

Huang, B., Bates, M., & Zhuang, X. W. (2009). Super-resolution fluorescence microscopy. *Annual Review of Biochemistry, 78*, 993–1016.

Hung, A. M., Micheel, C. M., Bozano, L. D., Osterbur, L. W., Wallraff, G. M., & Cha, J. N. (2010). Large-area spatially ordered arrays of gold nanoparticles directed by lithographically confined DNA origami. *Nature Nanotechnology, 5*, 121–126.

Jahn, K., Torring, T., Voigt, N. V., Sorensen, R. S., Kodal, A. L. B., Andersen, E. S., Gothelf, K. V., & Kjems, J. (2011). Functional patterning of DNA origami by parallel enzymatic modification. *Bioconjugate Chemistry, 22*(4), 819–823.

Jungmann, R., Liedl, T., Sobey, T. L., Shih, W., & Simmel, F. C. (2008). Isothermal assembly of DNA origami structures using denaturing agents. *Journal of the American Chemical Society, 130*, 10062–10063.

Jungmann, R., Steinhauer, C., Scheible, M., Kuzyk, A., Tinnefeld, P., & Simmel, F. C. (2010). Single-molecule kinetics and super-resolution microscopy by fluorescence imaging of transient binding on DNA origami. *Nano Letters, 10*, 4756–4761.

Kato, T., Goodman, R. P., Erben, C. M., Turberfield, A. J., & Namba, K. (2009). High-resolution structural analysis of a DNA nanostructure by cryoEM. *Nano Letters, 9*, 2747–2750.

Ke, Y. G., Lindsay, S., Chang, Y., Liu, Y., & Yan, H. (2008). Self-assembled water-soluble nucleic acid probe tiles for label-free RNA hybridization assays. *Science, 319*, 180–183.

Kuzyk, A., Laitinen, K. T., & Torma, P. (2009). DNA origami as a nanoscale template for protein assembly. *Nanotechnology, 20*(23), 235305.

Liedl, T., Hogberg, B., Tytell, J., Ingber, D. E., & Shih, W. M. (2010). Self-assembly of three-dimensional prestressed tensegrity structures from DNA. *Nature Nanotechnology, 5*, 520–524.

Pal, S., Deng, Z., Ding, B., Yan, H., & Liu, Y. (2010). DNA-origami-directed self-assembly of discrete silver-nanoparticle architectures. *Angewandte Chemie, International Edition, 49*(15), 2700–2704.

Rothemund, P. W. K. (2006). Folding DNA to create nanoscale shapes and patterns. *Nature, 440*, 297–302.

Rust, M. J., Bates, M., & Zhuang, X. (2006). Sub-diffraction-limit imaging by stochastic optical reconstruction microscopy (STORM). *Nature Methods, 3*, 793–795.

Sacca, B., Meyer, R., Erkelenz, M., Kiko, K., Arndt, A., Schroeder, H., Rabe, K., & Niemeyer, C. M. (2010). Orthogonal protein decoration of DNA origami. *Angewandte Chemie, International Edition, 49*, 9378–9383.

Schermelleh, L., Heintzmann, R., & Leonhardt, H. (2010). A guide to super-resolution fluorescence microscopy. *The Journal of Cell Biology, 190*, 165–175.

Seeman, N. C. (1982). Nucleic acid junctions and lattices. *Journal of Theoretical Biology, 99*, 237–240.

Selmi, D. N., Adamson, R. J., Attrill, H., Goddard, A. D., Gilbert, R. J. C., Watts, A., & Turberfield, A. J. (2011). DNA-templated protein arrays for single-molecule imaging. *Nano Letters, 11*, 657–660.

Sharonov, A., & Hochstrasser, R. M. (2006). Wide-field subdiffraction imaging by accumulated binding of diffusing probes. *Proceedings of the National Academy of Sciences of the United States of America, 103*, 18911–18916.

Sobey, T. L., Renner, S., & Simmel, F. C. (2009). Assembly and melting of DNA nanotubes from single-sequence tiles. *Journal of Physics-Condensed Matter, 21*, art. no. 034112.

Steinhauer, C., Jungmann, R., Sobey, T., Simmel, F., & Tinnefeld, P. (2009). DNA origami as a nanoscopic ruler for super-resolution microscopy. *Angewandte Chemie, International Edition, 48*, 8870–8873.

Vogelsang, J., Kasper, R., Steinhauer, C., Person, B., Heilemann, M., Sauer, M., & Tinnefeld, P. (2008). A reducing and oxidizing system minimizes photobleaching and blinking of fluorescent dyes. *Angewandte Chemie, International Edition, 47*, 5465–5469.

Vogelsang, J., Steinhauer, C., Forthmann, C., Stein, I. H., Person-Skegro, B., Cordes, T., & Tinnefeld, P. (2010). Make them blink: Probes for super-resolution microscopy. *Chemphyschem, 11*, 2475–2490.

Winfree, E., Liu, F. R., Wenzler, L. A., & Seeman, N. C. (1998). Design and self-assembly of two-dimensional DNA crystals. *Nature, 394*, 539–544.

Zhang, Y. W., & Seeman, N. C. (1994). Construction of a DNA-truncated octahedron. *Journal of the American Chemical Society, 116*, 1661–1669.

Zheng, J., Birktoft, J. J., Chen, Y., Wang, T., Sha, R., Constantinou, P. E., Ginell, S. L., Mao, C., & Seeman, N. C. (2009). From molecular to macroscopic via the rational design of a self-assembled 3D DNA crystal. *Nature, 461*, 74–77.

Chapter 10
DNA Nanotechnology

Ofer I. Wilner, Bilha Willner, and Itamar Willner

Abstract The base sequence encoded in nucleic acids yields significant structural and functional properties into the biopolymer. The resulting nucleic acid nanostructures provide the basis for the rapidly developing area of DNA nanotechnology. Advances in this field will be exemplified by discussing the following topics: (i) Hemin/G-quadruplex DNA nanostructures exhibit unique electrocatalytic, chemiluminescence and photophysical properties. Their integration with electrode surfaces or semiconductor quantum dots enables the development of new electrochemical or optical bioanalytical platforms for sensing DNA. (ii) The encoding of structural information into DNA enables the activation of autonomous replication processes that enable the ultrasensitive detection of DNA. (iii) By the appropriate design of DNA nanostructures, functional DNA machines, acting as "tweezers", "walkers" and "stepper" systems, can be tailored. (iv) The self-assembly of nucleic acid nanostructures (nanowires, strips, nanotubes) allows the programmed positioning of proteins on the DNA templates and the activation of enzyme cascades.

Keywords DNA • Nanotechnology • Sensors • Machines • Nanostructures

The base sequence in DNA encodes substantial structural and functional information into the biopolymer (Teller and Willner 2010). Besides the dictated duplex formation of complementary nucleic acids, the self assembly of single-stranded nucleic acids into G-quadruplex or i-motif (C-quadruplexes) (Phan and Mergny 2002) represent structural features of DNA. Similarly, the base sequence of nucleic acids dictates functional properties of the biopolymer, such as the selective scission by nicking enzymes or endonucleases, the sequence-specific binding of proteins, the specific binding of low-molecular-weight substrates or macromolecules (aptamers) (Ellington and Szostak 1990; Tuerk and Gold 1990; Mayer 2009) and the catalytic functions of nucleic acids (DNAzymes or ribozymes) (Breaker and Joyce 1994). These unique properties of DNA, together with the automated synthesis for the preparation of large quantities of DNA and modified nucleic acids by the polymerase chain reaction (PCR), provide a unique biomaterial that can be implemented for the development of functional nanostructures for different new applications, such as sensors, DNA-based machines, programmed DNA nanostructures and innovative applications for future nanomedicine (Bath and Turberfield 2007; Seeman 2007). The present article summarizes several recent advances of our laboratory in DNA nanotechnology.

O.I. Wilner • B. Willner • I. Willner (✉)
Institute of Chemistry, The Hebrew University of Jerusalem, Jerusalem 91904, Israel
e-mail: willnea@vms.huji.ac.il

E. Zahavy et al. (eds.), *Nano-Biotechnology for Biomedical and Diagnostic Research*,
Advances in Experimental Medicine and Biology 733, DOI 10.1007/978-94-007-2555-3_10,

10.1 DNA Nanostructures for Amplified Sensing

The analysis of DNA has important medical implications (detection of genetic disorders, or of pathogens, environmental food and agricultural analytical applications (analysis of bacteria), homeland security and forensic applications). The development of electronic or optical DNA sensing platforms attracted substantial efforts in the last two decades and several review articles addressed the advances in the field (Willner et al. 2008; Drummond et al. 2003). Nonetheless, challenging issues in developing ultrasensitive and specific DNA sensing systems are still ahead of us. The polymerase chain reaction (PCR) represents the "gold-label" in DNA analysis, yet it suffers from being error-prone and sensitive to contaminations. Furthermore, it requires instrumentational facilities for replication detection and method is accompanied by an intrinsic time-duration associated with thermal cycles of the replication. Also, the multiplexed analysis of samples and the quantitative analysis of the target DNA is difficult. Accordingly, the development of rapid, specific and quantitative DNA detection platforms with multiplexed analysis abilities that can be rapidly applied as field devices, is a long-term goal in analytical chemistry. Similarly, the discovery of the versatile systematic evolution of ligands by exponential enrichment (SELEX) method to select nucleic acid sequences with selective recognition properties for binding low-molecular weight substrates or macromolecules (aptamers) introduced new opportunities to assemble different nucleic acid-based sensor systems (Willner and Zayats 2007; Tombelli and Mascini 2009). The following section will describe several electrochemical and optical amplification schemes that provide scientific paradigms for solving the problems associated with DNA analysis.

Figure 10.1 a outlines the amplified electrochemical detection of DNA, or the development of aptasensors, using the hemin/G-quadruplex as amplifying electrocatalyst (Pelossof et al. 2010). The hemin/G-quadruplex (**1.1**) acts as a horseradish peroxidase mimicking DNAzyme, and different colorimetric or chemiluminescence DNA sensing assays implemented the DNAzyme as amplifying label. We found that this DNAzyme acts as electrocatalyst for the reduction of H_2O_2. This function was applied to develop DNA sensors. The hairpin nucleic acid structure (**1.2**) includes in the stem duplex region the G-quadruplex sequence in a "caged" inactive configuration, and the single-stranded loop domain acts as the recognition sequence for the analyte. Hybridization of the analyte (**1.3**) with the loop domain opens the hairpin nanostructure, resulting in the self-assembly of the hemin/G-quadruplex electrocatalyst that electrocatalyzes the reduction of H_2O_2. Figure 10.1b shows the electrocatalytic cathodic currents generated by the modified electrode in the presence of different concentrations of the target DNA, and the resulting calibration curve, Fig. 10.1b, inset. The method enabled the detection of DNA with a sensitivity that corresponded to 1×10^{-12} M. A generic aptasensor configuration was, similarly, developed using the hemin/G-quadruplex as amplifying electrocatalyst. This is exemplified in Fig. 10.1c with the development of an electrochemical aptasensor for AMP (**1.4**). The hairpin structure (**1.5**) assembled on a Au electrode included in the duplex stem region the G-quadruplex sequence in an inactive configuration, and the single stranded loop consisted of the anti-AMP sequence. In the presence of AMP the hairpin structure opened with the concomitant self-assembly of the hemin/G-quadruplex electrocatalytic nanostructure. Figure 10.1d depicts the electrocatalytic cathodic currents generated by the (**1.5**)-modified electrode in the presence of different concentration of AMP, and the resulting calibration curve. The system enabled the detection of AMP with a sensitivity that corresponded to 1×10^{-6} M.

The conjugation of the hemin/G-quadruplex to CdSe/ZnS quantum dots, QDs, enabled the development of optical DNA sensors or aptasensors (Sharon et al. 2010). The hemin/G-quadruplex exhibits a quasi-reversible redox wave at $E^o = -0.026$ V vs. NHE. This enables the electron transfer quenching of the photoexcited QDs, as schematically shown in Fig. 10.2a with the respective energy level diagram. Accordingly, the CdSe/ZnS QDs were functionalized with the hairpin nucleic acid (**1.6**) that included in the duplex stem region the G-quadruplex sequence in a protected inactive structure, while the loop region consisted of the sequence complementary to the target DNA, (**1.7**)

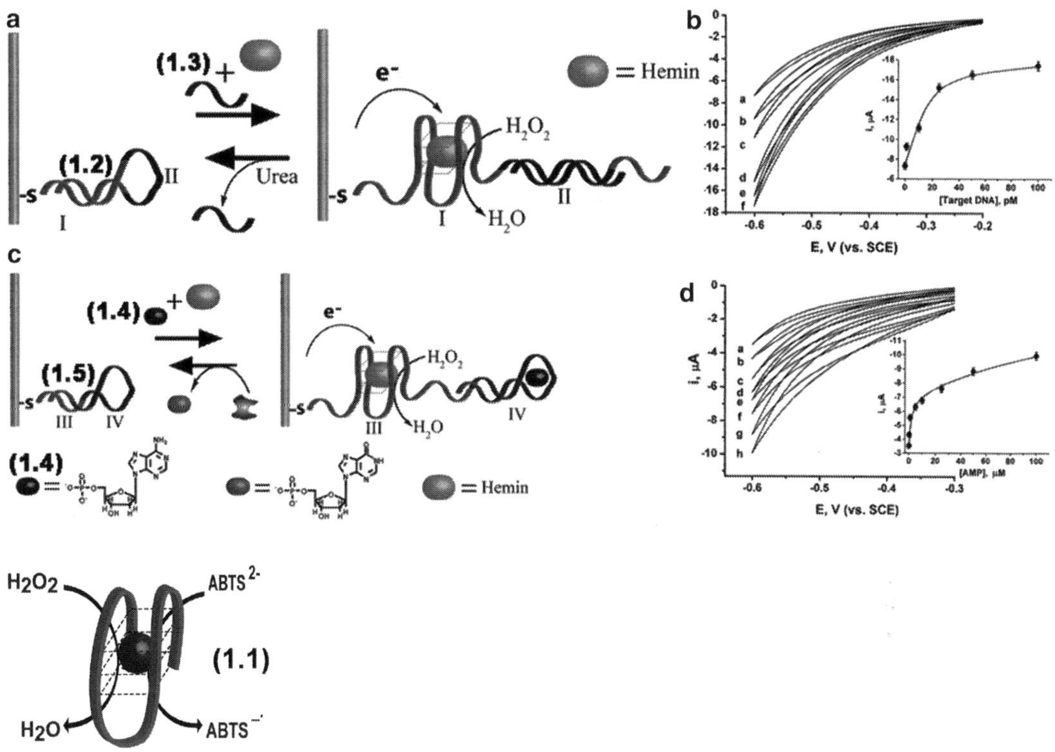

Fig. 10.1 (**a**) Sensing of a target DNA (**1.3**) by the opening of a thiolated hairpin nucleic acid structure (**1.2**) and the activation of a HRP-mimicking DNAzyme that electrolyzes the reduction of H_2O_2. (**b**) Electrocatalytic reduction of H_2O_2 by the (**1.3**)-activated HRP-mimicking DNAzyme caged in the DNA hairpin structure (**1.2**) linked to a Au electrode, in the presence of variable concentrations of (**1.3**): (*a*) 0, (*b*) 20, (*c*) 40, (*d*) 60, (*e*) 80, and (*f*) 100 pM. (**c**) Sensing of AMP (**1.4**) by the opening of the thiolated nucleic acid hairpin (**1.5**), associated with the electrode, and the formation of an aptamer–AMP Complex. (**d**) Electrocatalytic reduction of H_2O_2 by the (**1.4**)-activated HRP-mimicking DNAzyme caged in the nucleic acid hairpin structure (**1.5**) linked to a Au electrode, in the presence of different concentrations of (**1.4**): (*a*) 0, (*b*) 1, (*c*) 2, (*d*) 5, (*e*) 10, (*f*) 25, (*g*) 50, and (*h*) 100 μM (Reproduced with permission from Pelossof et al. 2010)

Fig. 10.2b. In the presence of the target, the hairpin nanostructure opened, and this resulted in the electron-transfer quenching of the luminescence of the QDs. Figure 10.2c shows the time-dependent decrease of the luminescence of the QDs in the presence of the analyte DNA, 1×10^{-6} M. The time-dependent changes in the luminescence represent the dynamics of opening of the hairpin nanostructure by the target DNA. As the kinetics of opening of the hairpin structure is controlled by the concentration of the analyte DNA, (**1.7**), probing the luminescence intensities of the QDs at a fixed time-interval of opening the hairpin structure by variable concentrations of (**1.7**) enabled the quantitative analysis of the target DNA. The resulting calibration curve is shown in Fig. 10.2d. The similar concept was used to develop optical aptasensors. This was achieved by caging the G-quadruplex sequence in the stem of the probe hairpin, and the incorporation of the aptamer recognition sequence in the single-stranded loop of the hairpin structure.

The hemin/G-quadruplex horseradish peroxidase mimicking DNAzyme reveals, also, chemiluminescence functions (Xiao et al. 2004; Freeman et al. 2011). It was found that the DNAzyme catalyzes the oxidation of luminol by H_2O_2, while generating chemiluminescence ($\lambda = 420$ nm). The light generated by the DNAzyme in proximity of the QDs, provided a "local light source" for the excitation of the QDs. The resulting chemiluminescence resonance energy transfer (CRET) process triggered on the luminescence of the QDs. This CRET process was used to develop optical aptasensors and DNA sensing systems and it was applied for the multiplexed parallel analysis of several DNA

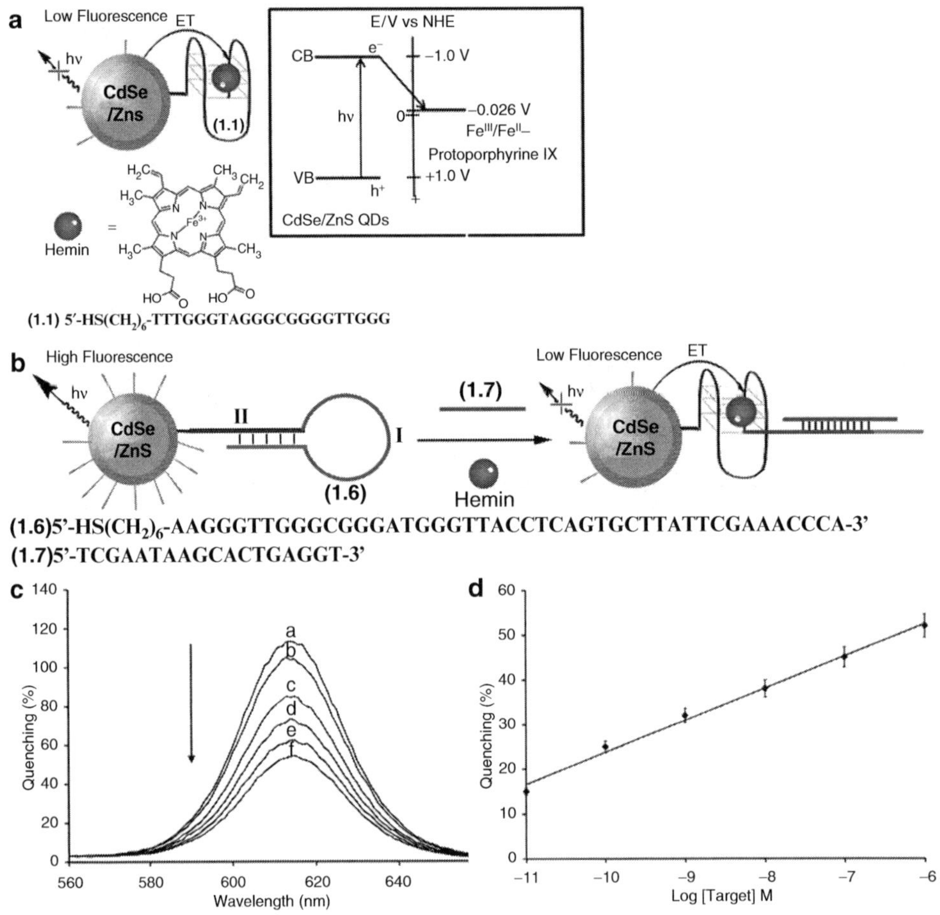

Fig. 10.2 (**a**) Self-assembly of QDs-G-qouadruolex/Hemin conjugates. (**b**) Schematic analysis of a target nucleic acid by G-Quadruplex-hairpin-modified QDs. (**c**) Time-dependent luminescence quenching of the (**1.1**)-modified QDs in the presence of hemin, 1×10^{-7} M, (*a*) before addition of target (**1.7**). (*b–f*) Upon interaction with target nucleic acid (**1.7**), 1×10^{-6} M. (**d**) Calibration curve corresponding to the luminescence quenching of the (**1.1**)/hemin-modified QDs at variable concentrations of (**1.7**) (Reproduced with permission from Sharon et al. 2010)

analytes. Figure 10.3a depicts schematically the CRET-based assembly of an ATP aptasensor. The nucleic acids (**1.8**) and (**1.9**) consist of the subunits of the G-quadruplex sequence (domains I and II) that are conjugated to the anti-ATP aptamer sequence subunits (domains III and IV). CdSe/ZnS QDs were functionalized with the nucleic acid subunit (**1.8**). While the two nucleic acid do not form a stable inter-sequence hemin/G-quadruplex structure, formation of the ATP/aptamer subunits complex stabilizes cooperatively the self-assembly of the nanostructure that includes the hemin/G-quadruplex adjacent to the CdSe/ZnS QDs. The chemiluminescence generated by the hemin/G-quadruplex stimulates the CRET process to the QDs, and the resulting luminescence of the QDs provides a readout signal for the ATP analyte. Figure 10.3b depicts the CRET-stimulated luminescence of the QDs generated by the formation of the ATP/aptamer subunits/QDs complex after a fixed time-interval of assembly and at variable concentrations of ATP. The respective calibration curve was extracted, Fig. 10.3c. The method enabled the detection of ATP with a sensitivity corresponding to 1×10^{-6} M. Figure 10.3d, (a) outlines the schematic CRET-based analysis of a DNA target. The CdSe/ZnS QDs were modified with the hairpin structure (**1.10**) that included in the stem region the caged G-quadruplex region and in the loop region the recognition sequence for detection of the analyte. In the presence of the analyte (**1.11**) the loop was opened, and this resulted in the self assembly

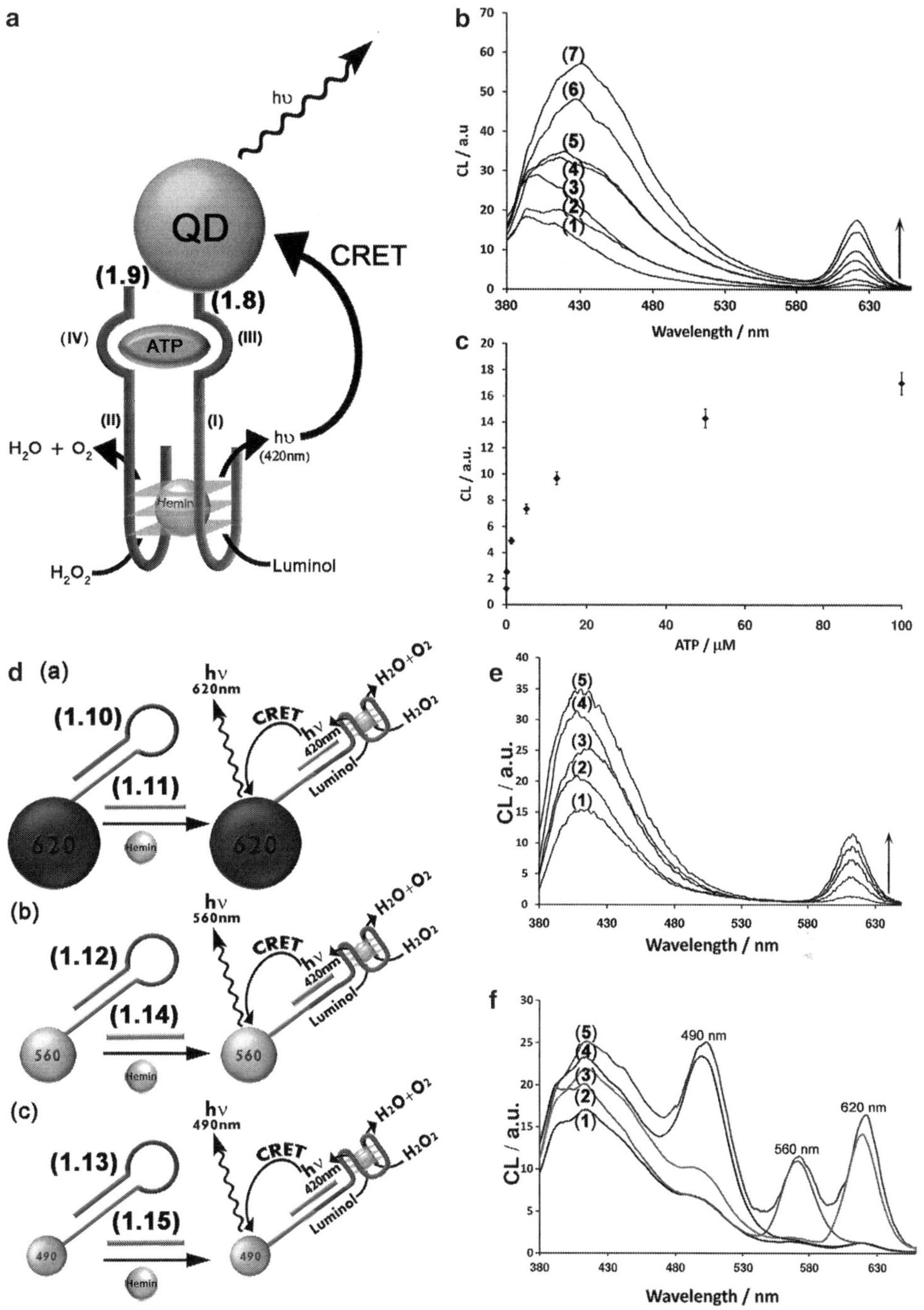

Fig. 10.3 (**a**) Analysis of ATP through the CRET from luminol, oxidized by the assembled hemin-G-quadruplex, to the QDs. (**b**) Luminescence spectrum corresponding to the CRET signal of the QDs at $\lambda = 612$ nm in the absence of ATP, curve (*1*) and in the presence of different concentrations of ATP: (*2*) 1.25×10^{-7} M (*3*) 1.25×10^{-6} M (*4*) 5×10^{-6} M (*5*) 12.5×10^{-6} M (*6*) 5×10^{-5} M (*7*) 1×10^{-4} M. (**c**) Calibration curve corresponding to the increase in the CRET signal at $\lambda = 620$ nm. (**d**) Multiplexed analysis of the different target DNAs using three different sized CdSe/ZnS QDs emitting at (*a*) 620 nm, (*b*) 560 nm and (*c*) 490 nm using the CRET sensing platform. (**e**) Luminescence spectrum corresponding to the CRET signal of the QDs at $\lambda = 620$ nm in the absence of (**1.11**), curve (*1*) and in the presence of different concentrations of (**1.11**): (*2*) 10 nM (*3*) 25 nM (*4*) 50 nM (*5*) 00 nM. (**f**) (*1*) Luminescence spectrum in the presence of the QDs mixture corresponding to the CRET signal in the absence of the different DNA targets. (*2*) In the presence of the QDs mixtures and (**1.14**). (*3*) In the presence of the QDs mixture and (**1.11**). (*4*) In the presence of the QDs mixture and (**1.15**). (*5*) In the presence of the QDs mixture and all three targets

of the hemin/G-quadruplex. The resulting DNAzyme catalyzed the generation of chemiluminescence that affected a CRET process to the QDs, resulting in their luminescence. Figure 10.3e shows the luminescence spectra of the QDs upon analyzing different concentrations of the target DNA, (**1.11**). Figure 10.3f depicts the multiplexed, parallel, analysis of three different DNA targets, using three different-sized QDs. While the three different sized QDs can be excited by the same chemiluminescence light source the resulting luminescence emissions are controlled by the size of the QDs. Accordingly, three different sized QDs were functionalized with these different nucleic acid hairpin structures (**1.10**), (**1.12**) and (**1.13**). The stem region of all hairpins included the cages sequence of the G-quadruplex whereas each of the hairpins included a different single-stranded loop domain that was complementary to the targets (**1.11**), (**1.14**) and (**1.15**), respectively. In the presence of each of the DNA targets, each of the hairpins was selectively opened, resulting in the self-assembly of the hemin/G-quadruplex and the generation of chemiluminescence. As a result, the selective formation of the hybrid between the DNA target and the respective hairpin probe dictated the CRET-stimulated luminescence of the respective QDs, Fig. 10.3f. Similarly, by mixing any of the two DNA targets the luminescence of the respective QDs was activated, and by treatment of the mixture consisting of the three-sized QDs with all three DNA analytes the resolved luminescence of all three QDs was observed. Figure 10.3f, demonstrated the multiplexed analysis of nucleic acid targets. Thus, by implementing other sized QDs and different-sized QDs of other materials, one may envisage the application of the CRET-based multiplexed analysis of DNAs.

10.2 Ultrasensitive Detection of DNA Through Isothermal Replication Processes Using DNAzymes

The analysis of DNA using catalytic labels reveals limited amplification capacity that is controlled by the catalytic turnover of the catalyst. The development routes that replicate the catalytic label as a result of the recognition event, or eventually replicate the analyte itself upon recognition of the analyte, could add new dimensions to the ultrasensitive detection of DNA. In the present section we implement the functional information encoded in the base sequence of nucleic acids, and the catalytic properties of nucleic acids (DNAzymes) to design sensitive isothermal DNA sensing platforms that could eventually substitute the legendary PCR method.

Figure 10.4a outlines the development of an analyte-triggered autonomous replication of the DNAzyme label that transduces optically the recognition event. The sensing platform consists of an engineered nucleic acid track (**2.1**) acting as the functional unit for the isothermal replication device of the DNAzyme label (Weizmann et al. 2006). The track (**2.1**) includes three functional domains; Domain I acts as the recognition site of the analyte nucleic acid, whereas Domain III is composed of the sequence complementary to the horseradish peroxidase hemin/G-quadruplex DNAzyme. The domain II is considered as the "heart" of the molecular device, and upto its replication yields a complementary sequence designed to be cleaved by a nicking enzyme, e.g., N.BbvCI. Thus, the sensor system consists of the track (**2.1**), the enzymes polymerase, the nicking enzyme N.BbvCI, and the nucleotide mixture, dNTPs. The analyte (**2.2**) is recognized by domain I of the track and results in the recognition complex. This triggers on the replication of the track, while generating the nicking domain in the replicated strand. The nicking of the resulting strand by N.BbvCI yields an open site for the replication of the track, resulting in the strand displacement of the replicated strand generated in the first cycle. The displaced strand consists, however, of the base sequence that self-assembles in the presence of hemin, into the hemin/G-quadruplex horseradish peroxidase mimicking DNAzyme. The generated DNAzyme acts as a catalytic label that amplifies the primary recognition event through the oxidation of 2′2′-azino-bis[3-ethylbenziazoline-6-sulfonic-acid], $ABTS^{2-}$ by H_2O_2 to the colored $ABTS^{\cdot-}$ product, $\lambda = 415$ nm, or by generation of chemiluminescence by the catalyzed oxidation of

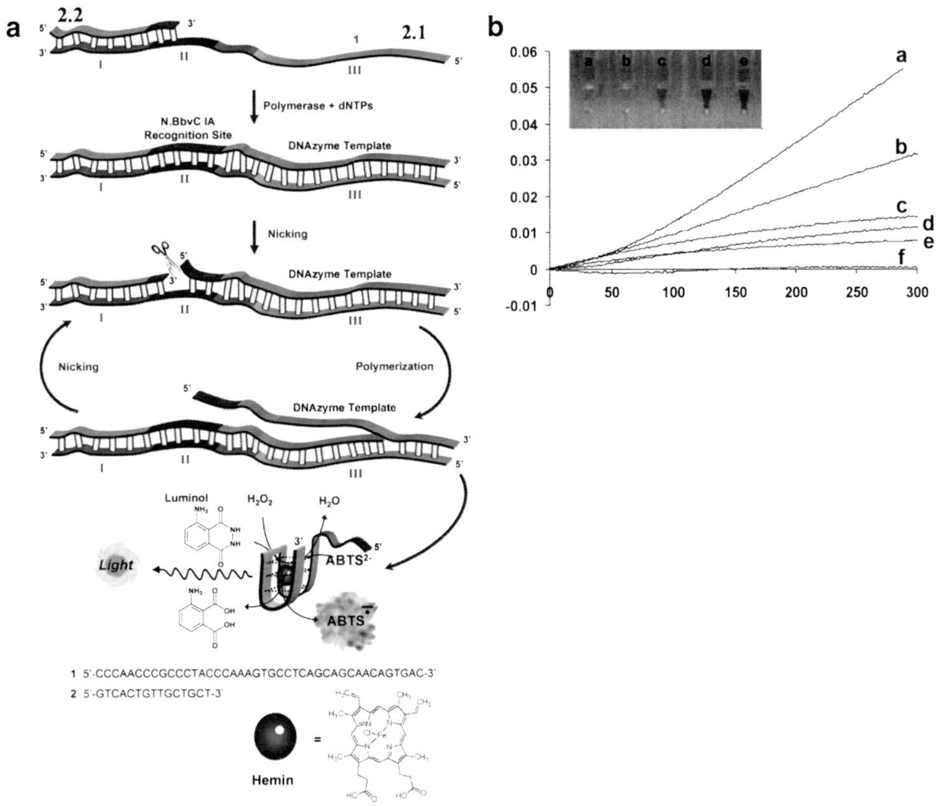

Fig. 10.4 (**a**) Primer-induced autonomous synthesis of DNAzyme units on template DNA by using polymerase/dNTPs and a nicking enzyme as the biocatalyst. (**b**) Absorbance changes observed upon the oxidation of $ABTS^{2-}$by the DNA-based machine (Reproduced with permission from Weizmann et al. 2006. Copyright Wiely-VCH Verlag GmbH & Co. KGaA)

luminol by H_2O_2 ($\lambda=420$ nm). Thus, the recognition of the analyte (**2.2**) by the track (**2.1**) activates the autonomous isothermal replication of the hemin/G-quadruplex DNAzyme, using the dNTPs nucleotide mixture as the "fuel" that generates the DNAzyme label as the "waste product". Figure 10.4b shows the time-dependent absorbance changes of the system upon analyzing different concentrations of the target DNA, (**2.2**). The detection of the analyte can be even visually analyzed, Fig. 10.4b, inset. Similarly, the chemiluminescence light intensities generated by the system in the presence of variable concentrations of (**2.2**) were analyzed. This method enables the analysis of DNA with a detection limit that corresponded to 1×10^{-14} M. While an apparent limitation of the sensing platform seems to be the need to design and optimize a track for each DNA target, this difficulty can be resolved by designing different hairpin probes that upon opening by the respective analytes activate a common DNA track replication machine (Weizmann et al. 2006).

The replication of the analyte by a DNAzyme-stimulated process is exemplified in Fig. 10.5a using the Mg^{2+}-dependent DNAzyme as a catalyst (Wang et al. 2011a). This DNAzyme cleaves a specific DNA sequence that includes a ribonucleo base as a scission site. The system included the nucleic acids (**2.3**) and (**2.4**) as functional units and the hairpin structure (**2.6**) as a substrate. The domains I and II of the nucleic acids (**2.3**) and (**2.4**) consist of the Mg^{2+}-dependent DNAzyme subunits. The subunit (**2.4**) exists in a hairpin configuration, where the single-stranded loop acts as a recognition site for the analyte DNA, (**2.5**), and the stem duplex region "cages" a part of the DNAzyme subunit (domain II). The caging of the catalytic subunits prevents the assembly of the Mg^{2+}-dependent

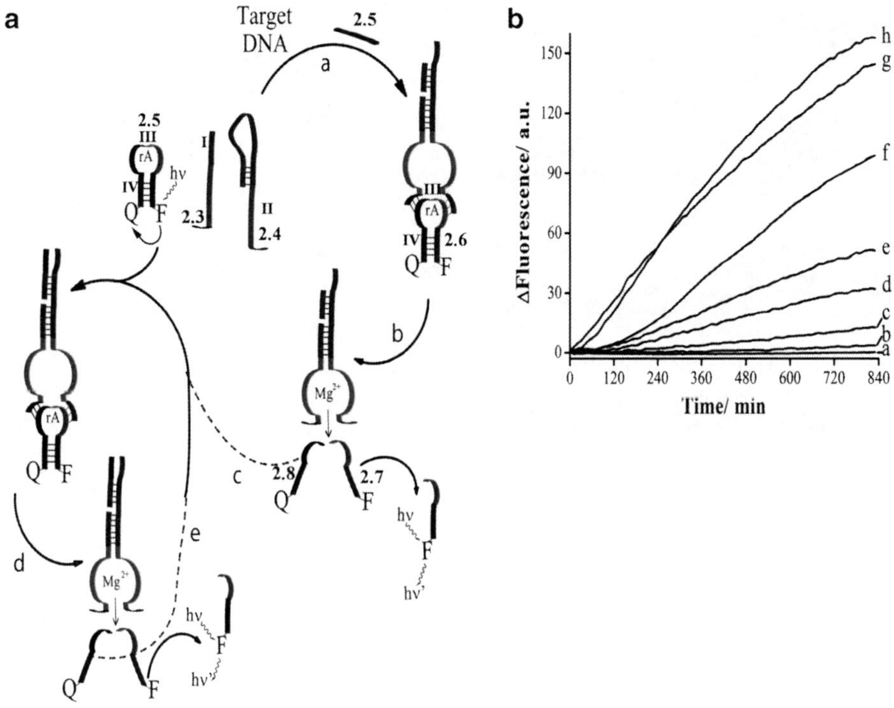

Fig. 10.5 (**a**) Schematic representation of the analyte-induced DNAzyme assembly and the sensing process: (*a*) target recognition and assembly of the DNAzyme, (*b*) cleavage of the substrate **2.6** and release of the activator unit **2.8**, (*c–e*) autocatalytic catabolic generation cycle triggered by the activator-unit-induced assembly of the DNAzyme. (**b**) Time-dependent fluorescence changes of the autocatalytic nucleic acid sensing in the presence of different target concentrations: (*a*) 0 M, (*b*) 1×10^{-12} M, (*c*) 1×10^{-11} M, (*d*) 1×10^{-10} M, (*e*) 1×10^{-9} M, (*f*) 1×10^{-8} M, (*g*) 1×10^{-7} M, and (*h*) 1×10^{-6} M. The initial fluorescence value at $t=0$ min was subtracted from the curves (Reproduced with permission from Wang et al. 2011a. Copyright Wiely-VCH Verlag GmbH & Co. KGaA)

DNAzyme. The substrate, (**2.6**), exists in a hairpin structure, where the single strand domain III includes the ribonucleobase-containing specific sequence for cleavage by the Mg^{2+}-dependent DNAzyme and the sequence IV consists of the sequence of the analyte (**2.5**). The 3′- and 5′ ends of the hairpin (**2.6**) are functionalized by a fluorophore-quencher pair. The intimate contact of the fluorophore-quencher pair yield the effective quenching of the fluorophore. In the presence of the analyte nucleic acid, (**2.5**), the autocatalytic, catabolic, regeneration of the analyte proceeds in the system, while transducing the recognition event by a fluores cence signal; The analyte opens the hairpin (**2.4**), resulting in the self-assembly of the Mg^{2+}-dependent DNAzyme, and cleavage of the substrate to subunits (**2.7**) and (**2.8**), that are separated from the catalytic structure. While the released fluorophore-labeled nucleic acid, (**2.7**), provides the fluorescence readout signal, the fragmented nucleic acid (**2.8**) consists of the analyte sequence. Thus, replication of the analyte enhances the opening of the hairpin (**2.4**) and this drives the autonomous autocatalytic generation of the fluorescence signal. Figure 10.5b depicts the time-dependent fluorescence changes in the sensing systems upon analyzing different concentrations of the analyte DNA. This autocatalytic amplification method enabled the detection of the analyte DNA with a sensitivity that corresponded to 1×10^{-12} M. The successful analysis of the target DNA by the autocatalytic system rests on the delicate balance of the nature and number of bases that prohibit the activation of the autocatalytic process in the absence of the analyte. This would suggest that for each analyte the composition of the components of the autocatalysis should be optimized, an obvious drawback of the sensing platform. This difficulty was resolved, however, by introducing an auxiliary hairpin structure that includes a variable single stranded loop for any analyte DNA and a constant caged sequence in the stem, that upon opening of the hairpin activates the autocatalytic device (Wang et al. 2011a).

10.3 Programmed Nanostructures Acting as DNA Machines

The information encoded in nucleic acids enables the construction of complex nanostructures. By the appropriate design of the systems programmed functions can be tailored. Recent developments in the area of DNA nanotechnology have implemented the information encoded in the nucleic acid biopolymers to construct systems exhibiting mechanical functions (Dittmer et al. 2004; Beissenhirtz and Willner 2006; Bath and Turberfield 2007). Different DNA-machines were constructed in the past few years, and these include the construction of DNA "tweezers" (Yurke et al. 2000; Han et al. 2008), "walkers" (Shih and Pierce 2004), motors (Bath et al. 2009) and more (Buranachai et al. 2006). In these systems the moving elements consist of nucleic acids, and their dictated movement is triggered by other DNA sequences that stimulate strand-displacement processes. While the area of DNA machines is at its infancy, it holds great promises as future molecular mechanical devices for programmed signal-triggered release of nucleic acids. Such systems may be used for the silencing of genes, the release of aptamers, for the selective inhibition of enzymes and for the multiplexed sensing of DNA sequences (Simmel 2007). One major challenge in the future development of DNA machines is the need for new triggering signals, such as metal ions or pH, that could act as input triggers for future nanomedical applications.

We have used a pH stimuli and Hg^{2+} ions as eternal triggers for the activation of DNA machines (Elbaz et al. 2009; Wang et al. 2010). Figure 10.6a shows the schematic activation of DNA tweezers by a pH signal. The tweezer construct consists of two arms (**3.1**) and (**3.2**) bridged by the nucleic acid sequence (**3.3**). The arms are further bridged by the nucleic acid (**3.4**) that retains the nanostructure in a closed configuration. The arms (**3.1**) and (**3.2**) include pre-designed domains I and II that are rich with cytosine bases. At acidic pH, pH = 5.2, these sequences self-assemble into i-motif, C-quadruplexes that reveal enhanced stability as compared to the duplex domains between the arms and the bridging units (**3.4**). As a result, the tweezers nanostructure is opened, and the strand (**3.4**) is released. Upon the switching of the pH of the system to pH = 7.2, the C-quadruplex structures are dissociated, resulting in the uptake of the strand (**3.4**) by the tweezers and their closure. Thus, by the switching of the pH of the system between pH = 7.2 and pH = 5.2 the tweezers nanostructure is reversibly transformed between "closed" and "open" states. The arm (**3.3**) was labeled at its 3′ and 5′-ends with a fluorophore/quencher pair. In the "closed" structure of the tweezers, the close distance separating the fluorophore/quencher pair, resulted in effective fluorescence resonance energy transfer (FRET) quenching of the fluorophore. At pH = 5.2, the tweezers existed in the "open" state, where the fluorophore/quencher pair was spatially separated, leading to inefficient FRET quenching of the fluorophore. Figure 10.6b shows that the fluorescence intensities of the system are, indeed, switched by a low value (at pH = 7.2) and high value (at pH = 5.2), and the cyclic opening and closure of the tweezers nanostructure is demonstrated, Fig. 10.6b, inset.

Figure 10.6c depicts the Hg^{2+}-ion triggered closure and opening of a tweezers nanostructure. The "arms" of the tweezers consist of the nucleic acids (**3.5**) and (**3.6**) and these are bridged by the nucleic acid (**3.7**). Since domains III and IV in the arms (**3.5**) and (**3.6**) include partial complementarity to the bridging nucleic acid (**3.8**), the domains III and IV, as well as the bridging unit (**3.8**), include non-complementary "mirror-positioned" thymine (T) bases, and these bases can generate T-Hg^{2+}-T complexes. While the partial complementarity between the "arms" (**3.5**) and (**3.6**) with (**3.8**) is insufficient to generate a stable "closed" structure of the tweezers, the formation of the T-Hg^{2+}-T complexes, in the presence of added Hg^{2+} ions, stabilizes cooperatively the formation of the (**3.8**)-bridged closed nanostructure of the tweezers. The addition of cysteine to the system results in the dissociation of the

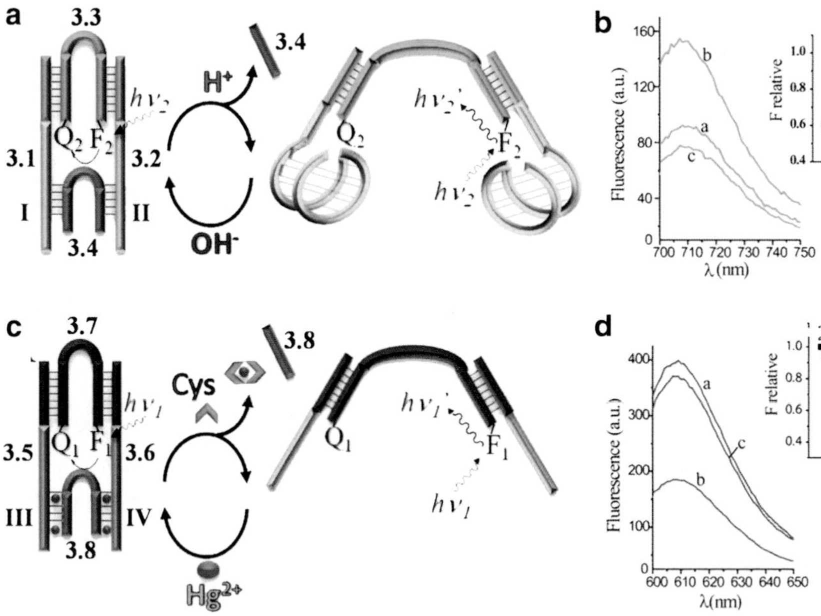

Fig. 10.6 (**a**) Schematic presentation of the opening and closing of pH dependent DNA tweezers using H^+ and OH^- as inputs that form the configurations "0" (closed) and "1" (opened), respectively. (**b**) Fluorescence spectra corresponding to the respective configurations of tweezers: (*a*) Initial state of the tweezers, (0), with no added input at pH = 7.2. (*b*) After the addition of H^+ as input that changes the pH value to 5.2. (*c*) After the addition of the OH^- input. Inset: Cyclic activation of the tweezers between configurations (1) (*b*) and (0) (*c*). (**c**) Schematic presentation of the concurrent activation of Hg^+ dependent DNA tweezers using Hg^{2+} and cysteine as inputs that form the configurations "0" (closed) and "1" (opened), respectively. (**d**) Fluorescence spectra corresponding to the respective configuration of tweezers: (*a*) Initial configuration of the tweezers, (1), with no added input. (*b*) After the addition of Hg^{2+} as input. (*c*) After the addition of the cysteine input. Inset: Cyclic activation of the tweezers between configurations (0, closed) (*b*) and (1, open) (*c*) (Reproduced with permission from Wang et al. 2010)

T-Hg^{2+}-T complexes due to the enhanced stability of the Hg^{2+}-cysteine complex. This leads to the release of (**3.8**), and to the opening of the tweezers nanostructure. The opening and closure of the tweezers was, similarly, followed by labeling the bridging unit (**3.7**) with a fluorophore/quencher pair. Figure 10.6d depicts the switchable fluorescence intensities of the tweezers nanostructure system in the presence of Hg^{2+} ions. In the absence of free Hg^{2+}-ions, the system existed in the open configuration that led to high fluorescence of the fluorophore label, curve (a). The closure of the tweezers, by added Hg^{2+} results in an intimate distance between the fluorophore/quencher pair, and this led to a low fluorescence intensity of the fluorescence label. Further addition of cysteine removed the (**3.8**) from the nanostructure and the resulted in the transformation of the tweezers into an open state, reflected by the high fluorescence intensity of the fluorophore. By the cyclic treatment of the system with Hg^{2+} ions and their removal by cysteine, the nanostructure was switched between "closed" and "open" states, respectively, Fig. 10.6d, inset.

The successful mechanical activation of DNA devices by pH or Hg^{2+}-ions as triggering stimuli was, then, implemented to design DNA machines of enhanced complexity. This is exemplified in Fig. 10.7a with the development of a "bi-pedal DNA-walker" (Wang et al. 2011b). A DNA construct consisting of four inter-hybridized footholds (**3.9**), (**3.10**), (**3.11**) and (**3.12**) acted as the "walker" track. Each of the footholds was modified with functional nucleic acids (**3.13**), (**3.14**), (**3.15**) and (**3.16**) that included the instructive information for the dictated "walkover" on the "track" of the "walker" element, consisted of the nucleic acids (**3.17**) and (**3.18**) that are bridged by the nucleic acid (**3.19**). The "walker" included two arms that may interact upon the application of appropriate triggers

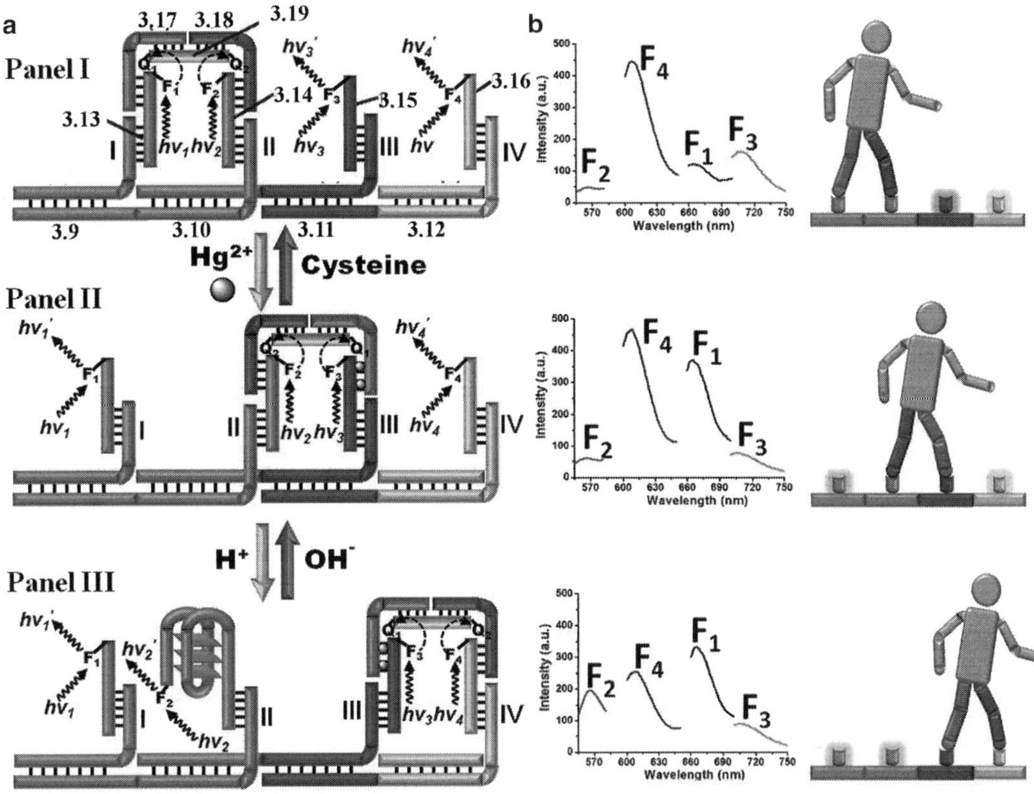

Fig. 10.7 (**a**) Schematic representation of the bipedal walker activated by H^+/OH^- and Hg^{2+}/cysteine inputs: *panel I*, walker immobilized on footholds I and II, imaged by high fluorescence of F_3 and F_4; *panel II*, walker positioned on footholds II and III imaged by fluorescence of F_1 and F_4; *panel III*, walker positioned on footholds III and IV, imaged by high fluorescence of F_1 and F_2. (**b**) time-dependent fluorescence changes upon the reversible activation of the bi-pedal walker (Printed with permission from Wang et al. 2011b. Copyright (2011) American Chemical Society)

with four different footholds. To follow the bi-pedal walker functions of the construct, the nucleic acids associated with the footholds (**3.13**), (**3.14**), (**3.15**) and (**3.16**) were functionalized with the fluorophore F_1, F_2, F_3 and F_4, whereas the bridging walker units (**3.19**) was functionalized at its 3′ and 5′ ends with black hole quencher units. Resting of the walker on any two footholds was then monitored by the quenching of the respective two fluorophores on the adjacent footholds. The "walker" system was initiated in the configuration outlined in Fig. 10.7a, panel I, where the arms of the walker are hybridized to footholds (**3.13**) and (**3.14**) to form the energetically-favored duplexes. This state of the walker construct is evident by the quenched fluorescence of F_1 and F_2, and high fluorescence intensities of the fluorophore F_3 and F_4. Addition of Hg^{2+} to the system drives the arm (**3.17**) to foothold (**3.15**) to form a duplex structure of enhanced stability due to cooperative base-pairing and the formation of T-Hg^{2+}-T complexes, Fig. 10.7a, panel II. This configuration of the walker is reflected by quenched fluorescence of F_2 and F_3 and high fluorescence intensities of F_1 and F_4. Addition of cysteine removes the Hg^{2+} from the duplex on the foothold (**3.15**), resulting in the backward "walking" of the pedal to foothold (**3.13**). The subsequent treatment of the walker system shown in panel II at pH=5.2 leads to the self-assembly of the functional nucleic acid (**3.14**) associated with foothold (**3.10**) into the i-motif, C-quadruplex. This leads to the walkover of the pedal (**3.18**), to the nucleic acid (**3.16**) associated with foothold (**3.12**), Fig. 10.7a, panel III. This is reflected by the efficient quenching of fluorophores F_3 and F_4 and high fluorescence intensities of F_1 and F_2. Naturalization of the system shown in panel III, pH = 7.0, dissociates the i-motif, and this stimulates the back walking of pedal (**3.18**) to form the energetically favored duplex between pedal (**3.18**) and the nucleic acid (**3.14**), associated with foothold (**3.10**). Figure 10.7b depicts the time-dependent fluorescence changes upon the reversible activation of the bi-pedal walker.

10.4 Self-Assembly of Functional DNA-Protein Nanostructures

Numerous scientific efforts are directed, recently, to the use of the information encoded in nucleic acids to self-assemble one-dimensional, two-dimensional, and three-dimensional nanostructures (Lin et al. 2006; Aldaye et al. 2008; Seeman 2010). Methods to construct DNA nanostructures by the application of the rolling circle amplification (RCA) process (Zhao et al. 2006) or the self-assembly of 1D nanostructures by molecular or macromolecular "glues" were reported (Cheglakov et al. 2007). Similarly, DNA-tiles consisting of "sticky-ends" toehold complementary nucleic acids were used to self-assemble two-dimensional and three-dimensional DNA nanotubes nanostructures (Fu and Seeman 1993; LaBean et al. 2000; Mathieu et al. 2005; Sharma et al. 2009). Also, by the use of poly-functional nucleic acids self-organization of DNA into covalently-bonded sheets and their wrapping into nanotubes was reported (Wilner et al. 2010). Impressive DNA nanostructures were fabricated by the "stapling" of the long M13 phage DNA with many complementary nucleic acids acting as "staple units" to form programmed two-dimensional or three-dimensional "origami" nanostructures (Rothemund 2006; Dietz et al. 2009). While the self-organization of DNA nanostructures introduced an important facet into the field of DNA nanotechnology, the design of functional DNA nanostructures is a challenging area that should be addressed. Different functionalities of DNA nanostructures may be envisaged such as templates for catalytic or biocatalytic transformations, nanoreactors for controlled synthesis, nano-containers for slow release, templates for the fabrication of nano-devices, systems with programmed machinery and motility functions, and more (Lo et al. 2010; Andersen et al. 2009; Keren et al. 2003; He and Liu 2010).

One important system includes the self-organization of DNA-protein hybrid nanostructures. In such systems the DNA component might provide the instructive information to programme the topology of the proteins, and the resulting ensemble of proteins might then yield emerging functions as a result of their organization. Figure 10.8a depicts the programmed ordering of two proteins thrombin and lysozyme on a DNA template (Cheglakov et al. 2007). A circular DNA, (**4.1**), that includes two domains I and II, complementary to the anti-thrombin and anti-lysozyme aptamers. In the presence of the primer (**4.2**) the nucleotide mixture, dNTPs, and polymerase (phi 29), the rolling circle amplification (RCA) process was activated, leading to the formation of the RCA chains that include alternate sequences of the thrombin aptamer and lysozyme aptamer. In the presence of thrombin or lysozyme, its selective binding of the proteins to the RCA chains was observed and in the presence of the two proteins the dense formation of alternately positioned proteins was demonstrated. The selective assembly of the proteins on the RCA chains, and the positioning of the two proteins on the DNA template was imaged by AFM and by confocal microscopy, Fig. 10.8b, c.

The activation of enzyme cascades on DNA nanostructures was exemplified with the use of one-dimensional hexagon-type strips as organizing medium of the enzymes (Wilner et al. 2009). The two nucleic acids (**4.3**) and (**4.4**) include appropriate complementarities that enables the self-assembly of hexagon-type strips, Fig. 10.9a, structure I, whereas the four nucleic acids (**4.5**), (**4.6**), (**4.7**) and (**4.8**) include the appropriate complementarities to yield the four-hexagon strips, Fig. 10.9a, structure II. The resulting strips were imaged by AFM, and they revealed width corresponding to ca. 15 nm and ca. 30 nm, consistent with their geometrical features. The "hexagon" units at the edges of the strips included toehold nucleic acid tethers III and IV that enabled the secondary attachment of proteins. Glucose oxidase, GOx was functionalized with the nucleic acid (**4.9**), complementary to the domain III, whereas the enzyme horseradish peroxidase, HRP, was modified with the nucleic acid (**4.10**).

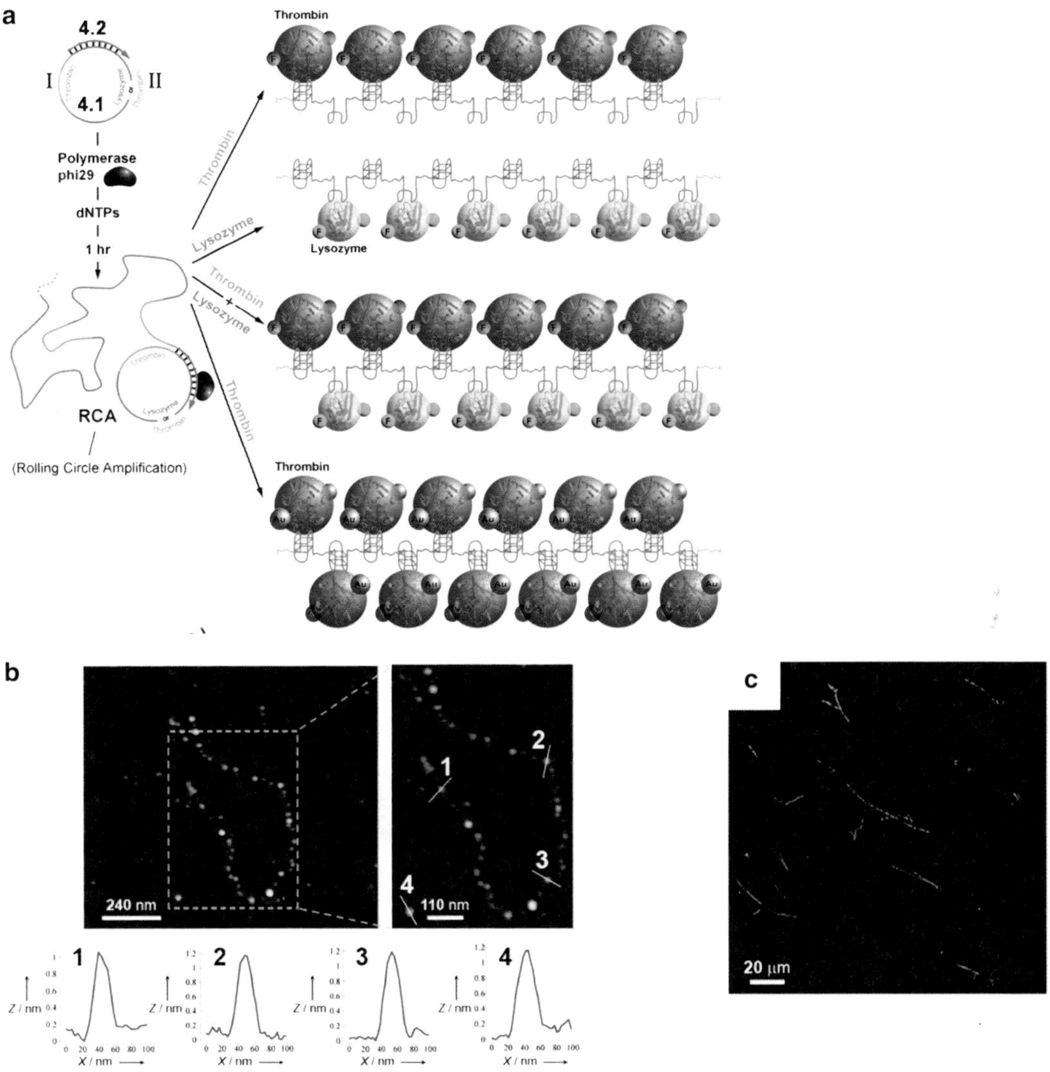

Fig. 10.8 (**a**) Preparation of periodic DNA tapes created using rolling-circle amplification. A circular template is subjected to polymerization to create the periodic tapes, which in turn can be reacted with the appropriate enzyme to form the protein-immobilized nanostructures. (**b**) AFM topographical image of thrombin-functionalized DNA tapes (**c**) CFM image of periodic DNA tapes modified with fluorescein-labeled lysozyme (Reproduced with permission from Cheglakov et al. 2007. Copyright Wiely-VCH Verlag GmbH & Co. KGaA)

These modified biocatalysts were then hybridized with the respective DNA strips, acting as templates for the programmed positioning of the enzymes on the DNA nanostructures, Fig. 10.10a. An enzyme cascade was then activated in the bi-enzyme/DNA hybrid nanostructures. Glucose oxidase catalyzed the O_2-mediated oxidation of glucose to gluconic acid and H_2O_2. The resulting H_2O_2 acted, however, as substrate for HRP that catalyzed the oxidation of 2′2′–azino-bis[3-ethylbenziazoline-6-sulfonic-acid], $ABTS^{2-}$, to the colored product $ABTS^{\cdot -}$, $\lambda = 414$ nm. The spatial proximitiy between the two enzymes that was dictated by the hexagonic DNA templates was found to control the activation of the bi-enzyme cascade, Fig. 10.10b; While the two enzymes did not communicate in a diffusional configuration of the two biocatalysts, or in the presence of a non-organizing DNA (calf thymus DNA), the organization of the enzymes on the DNA templates led to a topologically dictated concentration of the biocatalytic units. As a result, the H_2O_2 generated by the GOx-mediated oxidation of glucose is fed

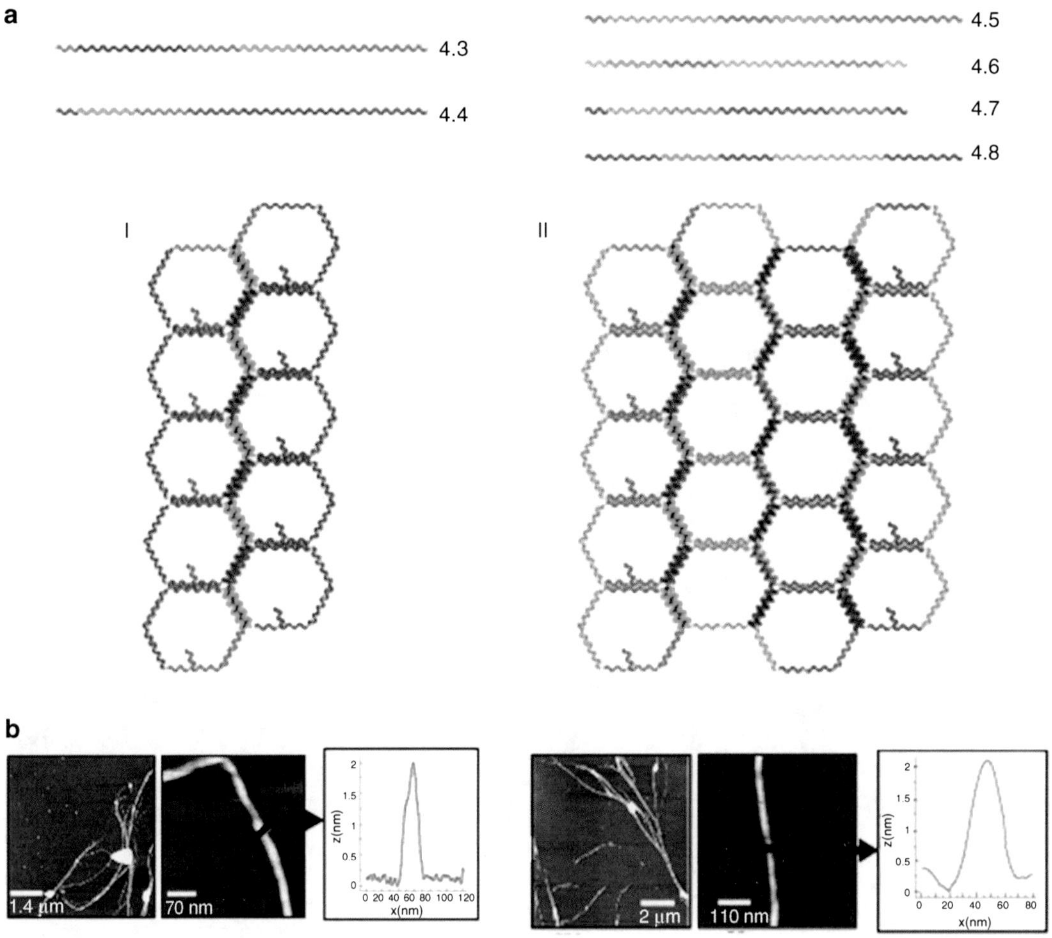

Fig. 10.9 (**a**) (I) Two-hexagon DNA strip assembly, (II) Four-hexagon DNA strip assembly. (**b**) AFM images and cross-section analysis of the two-hexagon and four-hexagon strips (Reproduced with permission from Wilner et al. 2009)

into the neighboring HRP biocatalyst that oxidizes $ABTS^{2-}$, Fig. 10.10b. Furthermore, it was demonstrated that the spatial separation of the enzymes by the DNA scaffold controlled the efficiency of the enzyme cascade, and the intimate contact between the enzymes in the two-hexagon configuration led to an enhanced activation of the cascade as compared to the four-hexagon DNA nanostructure.

The control of the biocatalytic activity of an enzyme/cofactor pair by the two-hexagon DNA nanostructure was demonstrated with the activation of the NAD^+-dependent glucose dehydrogenase. Glucose dehydrogenase, GDH, was modified with the nucleic acid (**4.11**) that is complementary to the toehold tether (III) of the two-hexagon strip. Aminoethyl nicotinamide adenenine di-nucleotide, H_2N-NAD^+ was functionalized with nucleic acids of variable lengths that are complementary at their ends to the toehold tether (IV) of the two-hexagon DNA template, Fig. 10.10c. The cofactor-mediated activation of GDH was demonstrated, where the GDH-catalyzed oxidation of glucose was accompanied by the reduction of the NAD^+ cofactor to NADH. The reduced NADH reduced methylene blue, MB^+, to the colorless product, MBH, thus providing a spectral signal for the process. It was found that the tether length associated with the NAD^+ cofactor to the template controls the biocatalytic transformation as a result of the spatial steric organization of the GDH/cofactor units on the DNA template, Fig. 10.10d. While the NAD^+ cofactor and GDH at identical concentrations in a diffusional state, or the NAD^+ cofactor and GDH in the presence of the foreign, non-organizing calf thymus DNA are not

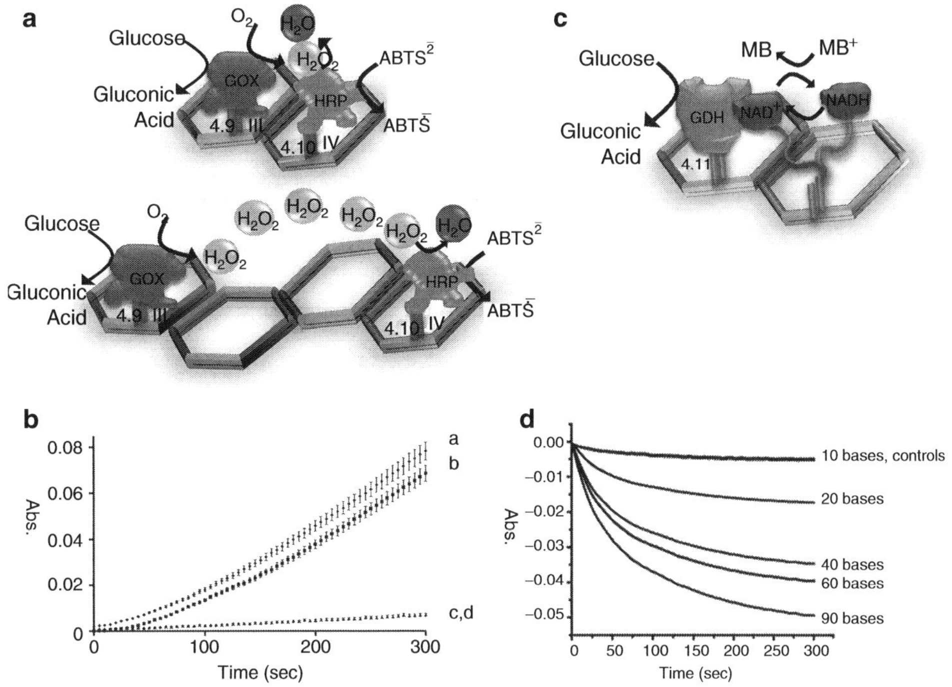

Fig. 10.10 (**a**) Assembly of the GOx and HRP enzymes on the two-hexagon and four-hexagon strips. (**b**) Time-dependent absorbance changes as a result of the oxidation of $ABTS^{2-}$ by the GOx–HRP cascade in the presence of (*a*) the two-hexagon scaffold, (*b*) the four-hexagon scaffold, (*c*) in the absence of any DNA, and (*d*) in the presence of foreign calf thymus DNA. (**c**) Assembly of the NAD^+/GDH system on the two-hexagon scaffold using different lengths of tethers linking the NAD^+ cofactor to the scaffold. (**d**) Time-dependent absorbance changes as a result of the cofactor-mediated oxidation of glucose by the 90-base chain tethered NAD^+ cofactor/GDH associated with the two-hexagon DNA scaffold, the 60-base chain tethered NAD^+ cofactor/GDH associated with the two-hexagon DNA scaffold, the 40-base chain tethered NAD^+ cofactor/GDH associated with the two-hexagon DNA scaffold, the 20-base chain tethered NAD^+ cofactor/GDH associated with the two-hexagon DNA scaffold, the 10-base chain tethered NAD^+ cofactor/GDH associated with the two-hexagon DNA scaffold, the long-chain and short-chain tethered NAD^+ cofactor/GDH systems in the presence of the foreign calf thymus DNA, respectively, the long-chain and short-chain tethered NAD^+ cofactor/GDH systems in the absence of DNA scaffold, respectively (Reproduced with permission from Wilner et al. 2009)

leading to the activation of the biocatalyst, the tether length linking the cofactor to the DNA template strongly affects the biocatalytic activation of GDH Fig. 10.10d. As the chain linking the NAD^+ cofactor to the two-hexagon DNA nanostructure is longer, the biocatalytic oxidation of glucose is enhanced. These results clearly indicate that the DNA template is essential to activate the coupled cofactor/enzyme biotransformation. The spatial concentration of the cofactor/enzyme units by the DNA scaffold and the intimate contact between the flexible, long chain-tethered, cofactor and the enzymes are essential to stimulate the biocatalytic oxidation of glucose.

10.5 Conclusions and Perspectives

DNA nanotechnology advanced in recent years to one of the most central research fields in modern science. The ease to synthesize nucleic acids of any base sequence and shape in large quantities, the possibility to select nucleic acid sequences of specific binding properties (aptamers) or catalytic functions (DNAzymes), and the ability to encode functional and structural information into the DNA biopolymers, provide a versatile material for the construction of functional nanostructures.

These different properties of DNA were implemented to develop a variety of research disciplines such as the use of DNA as active material for biocomputing and logic operations, the self-assembly of DNA into 1D, 2D and 3D nanostructures, and the development of new sensing platforms. Furthermore, nanotechnology provides nanoscale materials, such as nanoparticles, quantum dots, carbon nanotubes, metallic nanowires, and more that exhibit unique optical, electronic and catalytic properties. Not surprisingly, the conjugation of nucleic acids to different nanomaterials yields hybrid biomolecule/NPs (or carbon nanotubes, QDs, nanowires) structures that combine the recognition and catalytic properties of the biomolecules with the novel optical and electronic functions of the nanomaterials. Indeed, tremendous research efforts are directed in the last decade to synthesize biomoelcule-nanomaterial hybrid systems and to use them for optical or electronic sensing, imaging of cells, construction of nanoscale devices and nano-machines, nanomedical applications and more.

Our research laboratory pioneered the research in nanobiotechnology, and particularly research involving DNA nanotechnology. The present section highlighted some facets of our research in the field. One direction has addressed the use of metallic NPs or semiconductor QDs for the electronic and optical detection of DNA. While substantial progress was demonstrated in the application of metal nanoparticles and QDs for sensing, several challenging topics need further development. Particularly, the development of multiplex analysis platforms, for the detection of DNA, are needed. Here, the use of QDs, and the introduction of new photophysical mechanisms such as chemiluminescence resonance energy transfer (CRET) hold great promise. A further rapidly advancing research direction in DNA nanotechnology involves the use of DNA nanostructures as nanomachines. While these systems seem, at first glance, as a nano-engineering challenge, different future applications of such systems may be envisaged. The "mechanical" signal-triggered transport of nucleic acids might provide a versatile mechanism for autonomous nanomedicine, where the dictated release of nucleic acids might provide silencing sequences for genes (SiRNA) or specific inhibitors for enzymes (aptamers). Molecular DNA machines held, also, promise as functional nanostructures for amplified ultrasensitive sensing. The development of isothermal replication schemes or all-DNA autocatalytic replication processes, are anticipated to provide alternative routes to PCR for analyzing DNA. Also, the design of new configurations of machines is expected to enable new, innovative, functional nanostructures acting as carriers for controlled release or templates for dictated synthesis. Finally, substantial progress was demonstrated in the construction of DNA nanostructures. The ordered organization of proteins on the DNA templates may provide a general path to design signal transduction circuits and to activate enzyme cascades. Such protein/DNA hybrid systems could provide simple models for mimicking cellular processes and bridge the fields of system biology and system chemistry. While substantial progress was demonstrated in DNA nanotechnology, this interdisciplinary topic is still at its infancy, and exciting future discoveries may be expected in this field.

Acknowledgement Our research in DNA nanotechnology is supported by the Israel Science Foundation and the EC projects NANOGNOSTICS and ECCell.

References

Aldaye, F. A., Palmer, A. L., & Sleiman, H. F. (2008). Assembling materials with DNA as the guide. *Science, 321*, 1795–1799.

Andersen, E. S., Dong, M., Nielsen, M. M., et al. (2009). Self-assembly of a nanoscale DNA box with a controllable lid. *Nature, 459*, 73–76.

Bath, J., & Turberfield, A. J. (2007). DNA nanomachines. *Nature Nanotechnology, 2*, 275–284.

Bath, J., Green, S. J., Allen, K. E., & Turberfield, A. J. (2009). Mechanism for a directional, processive, and reversible DNA motor. *Small, 5*, 1513–1516.

Beissenhirtz, M. K., & Willner, I. (2006). DNA-based machines. *Organic and Biomolecular Chemistry, 4*, 3392–3401.

Breaker, R. R., & Joyce, G. F. (1994). A DNA enzyme that cleaves RNA. *Chemistry and Biology, 1*, 223–229.

Buranachai, C., Mckinney, S. A., & Ha, T. (2006). Single molecule nanometronome. *Nano Letters, 6*, 496–500.

Cheglakov, Z., Weizmann, Y., Braunschweig, A. B., Wilner, O. I., & Willner, I. (2007). Increasing the complexity of periodic protein nanostructures by the rolling-circle amplified synthesis of aptamers. *Angewandte Chemie (International ed. in English), 47*, 126–130.

Dietz, H., Douglas, S. M., & Shih, W. M. (2009). Folding DNA into twisted and curved nanoscale shapes. *Science, 325*, 725–730.

Dittmer, W. U., Reuter, A., & Simmel, F. C. (2004). A DNA-based machine that can cyclically bind and release thrombin. *Angewandte Chemie (International ed. in English), 43*, 3550–3553.

Drummond, T. G., Hill, M. G., & Barton, J. K. (2003). Electrochemical DNA sensors. *Nature Biotechnology, 21*, 1192–1199.

Elbaz, J., Wang, Z.-G., Orbach, R., & Willner, I. (2009). pH-stimulated concurrent mechanical activation of two DNA "tweezers". A "SET-RESET" logic gate system. *Nano Letters, 9*, 4510–4514.

Ellington, A. D., & Szostak, J. W. (1990). In vitro selection of RNA molecules that bind specific ligands. *Nature, 346*, 818–822.

Freeman, R., Liu, X., & Willner, I. (2011). Chemiluminescent and chemiluminescence resonance energy transfer (CRET) detection of DNA, metal ions and aptamer-substrate complexes using hemin/G-quadruplex and CdSe/ZnS quantum dots. *Journal of the American Chemical Society, 133*(30), 11597–11604.

Fu, T. J., & Seeman, N. C. (1993). Symmetric immobile DNA branched junctions. *Biochemistry, 32*, 8062–8067.

Han, X., Zhou, Z., Yang, F., & Deng, Z. (2008). Catch and release: DNA tweezers that can capture, hold, and release an object under control. *Journal of the American Chemical Society, 130*, 14414–14415.

He, Y., & Liu, D. R. (2010). Autonomous multistep organic synthesis in a single isothermal solution mediated by a DNA walker. *Nature Nanotechnology, 5*, 778–782.

Keren, K., Berman, R. S., Buchstab, E., Sivan, U., & Braun, E. (2003). DNA-templated carbon nanotube field-effect transistor. *Science, 302*, 1380–1382.

LaBean, T. H., Yan, H., Kopatsch, J., et al. (2000). Construction, analysis, ligation, and self-assembly of DNA triple crossover complexes. *Journal of the American Chemical Society, 122*, 1848–1860.

Lin, C., Liu, Y., Rinker, S., & Yan, H. (2006). DNA tile based self-assembly: Building complex nanoarchitectures. *ChemPhysChem, 7*, 1641–1647.

Lo, P. K., Karam, P., Aldaye, F. A., McLaughlin, C. K., Hamblin, G. D., Cosa, G., & Sleiman, H. F. (2010). Loading and selective release of cargo in DNA nanotubes with longitudinal variation. *Nature Chemistry, 2*, 319–328.

Mathieu, F., Liao, S., Kopatsch, J., Wang, T., Mao, C., & Seeman, N. C. (2005). Six-helix bundles designed from DNA. *Nano Letters, 5*, 661–665.

Mayer, G. (2009). The chemical biology of aptamers. *Angewandte Chemie (International ed. in English), 48*, 2672–2689.

Pelossof, G., Tel-Vered, R., Elbaz, J., & Willner, I. (2010). Amplified biosensing using the horseradish peroxidase-mimicking DNAzyme as an electrocatalyst. *Analytical Chemistry, 82*, 4396–4402.

Phan, A. T., & Mergny, J. (2002). Human telomeric DNA: G-quadruplex, i-motif and Watson-crick double helix. *Nucleic Acids Research, 30*, 4618–4625.

Rothemund, P. W. K. (2006). Folding DNA to create nanoscale shapes and patterns. *Nature, 440*, 297–302.

Seeman, N. C. (2007). An overview of structural DNA nanotechnology. *Molecular Biotechnology, 37*, 246–257.

Seeman, N. C. (2010). Structural DNA nanotechnology: Growing along with nano letters. *Nano Letters, 10*, 1971–1978.

Sharma, J., Chhabra, R., Cheng, A., Brownell, J., Liu, Y., & Yan, H. (2009). Control of self-assembly of DNA tubules through integration of gold nanoparticles. *Science, 323*, 112–116.

Sharon, E., Freeman, R., & Willner, I. (2010). CdSe/ZnS quantum dots-G-quadruplex/hemin hybrids as optical DNA sensors and aptasensors. *Analytical Chemistry, 82*, 7073–7077.

Shih, J., & Pierce, N. A. (2004). A synthetic DNA walker for molecular transport. *Journal of the American Chemical Society, 126*, 10834–10835.

Simmel, F. C. (2007). Towards biomedical applications for nucleic acid nanodevices. *Nanomedicine, 2*, 817–830.

Teller, C., & Willner, I. (2010). Functional nucleic acid nanostructures and DNA machines. *Current Opinion in Biotechnology, 21*, 376–391.

Tombelli, S., & Mascini, M. (2009). Aptamers as molecular tools for bioanalytical methods. *Current Opinion in Molecular Therapeutics, 11*, 179–188.

Tuerk, C., & Gold, L. (1990). Systematic evolution of ligands by exponential enrichment: RNA ligands to bacteriophage T4 DNA polymerase. *Science, 249*, 505–510.

Wang, Z.-G., Elbaz, J., Remacle, F., Levine, R. D., & Willner, I. (2010). All-DNA finite-state automata with finite memory. *Proceedings of the National Academy of Sciences of the United States of America, 107*, 21996–22001.

Wang, F., Elbaz, J., Teller, C., & Willner, I. (2011a). Amplified detection of DNA through an autocatalytic and catabolic DNAzyme-mediated process. *Angewandte Chemie (International ed. in English), 50*, 295–299.

Wang, Z.-G., Elbaz, J., & Willner, I. (2011b). DNA machines: Bipedal walker and stepper. *Nano Letters, 11*, 304–309.
Weizmann, Y., Beissenhirtz, M. K., Cheglakov, Z., Nowarski, R., Kotler, M., & Willner, I. (2006). A virus spotlighted by an autonomous DNA machine. *Angewandte Chemie (International ed. in English), 45*, 7384–7388.
Willner, I., & Zayats, M. (2007). Electronic aptamer-based sensors. *Angewandte Chemie (International ed. in English), 46*, 6408–6418.
Willner, I., Shlyahovsky, B., Zayats, M., & Willner, B. (2008). DNAzymes for sensing, nanobiotechnology and logic gate applications. *Chemical Society Reviews, 37*, 1153–1165.
Wilner, O. I., Weizmann, Y., Gill, R., Lioubashevski, O., Freeman, R., & Willner, I. (2009). Enzyme cascades activated on topologically programmed DNA scaffolds. *Nature Nanotechnology, 4*, 249–254.
Wilner, O. I., Henning, A., Shlyahovsky, B., & Willner, I. (2010). Covalently linked DNA nanotubes. *Nano Letters, 10*, 1458–1465.
Xiao, Y., Pavlov, V., Gill, R., & BourenkoT, W. I. (2004). Lighting up biochemiluminescence by the surface self-assembly of DNA-hemin complexes. *ChemBioChem, 5*, 374–379.
Yurke, B., Turberfield, A. J., Mills, A. P., Simmel, F. C., & Neumann, J. L. (2000). A DNA-fuelled molecular machine made of DNA. *Nature, 406*, 605–608.
Zhao, W., Gao, Y., Srinivas, A., et al. (2006). DNA polymerization on gold nanoparticles through rolling circle amplification: Towards novel scaffolds for three-dimensional periodic nanoassemblies. *Angewandte Chemie (International ed. in English), 45*, 2409–2413.

Chapter 11
Role of Carbohydrate Receptors in the Macrophage Uptake of Dextran-Coated Iron Oxide Nanoparticles*

Ying Chao, Priya Prakash Karmali, and Dmitri Simberg

Abstract Superparamagnetic iron oxide (SPIO, Ferumoxides, Feridex), an important MRI intravenous contrast reagent, is efficiently recognized and eliminated by macrophages in the liver, spleen, lymph nodes and atherosclerotic lesions. The receptors that recognize nanoparticles are poorly defined and understood. Since SPIO is coated with bacterial polysaccharide dextran, it is important to know whether carbohydrate recognition plays a role in nanoparticle uptake by macrophages. Lectin-like receptors CD206 (macrophage mannose receptor) and SIGNR1 were previously shown to mediate uptake of bacterial polysaccharides. We transiently expressed receptors MGL-1, SIGNR-1 and msDectin-1 in non-macrophage 293T cells using lipofection. The expression was confirmed by reverse transcription PCR. Following incubation with the nanoparticles, the uptake in receptor-expressing cells was not statistically different compared to control cells (GFP-transfected). At the same time, expression of scavenger receptor SR-A1 increased the uptake of nanoparticles three-fold compared to GFP-transfected and control vector-transfected cells. Blocking CD206 with anti-CD206 antibody or with the ligand mannan did not affect SPIO uptake by J774.A1 macrophages. Similarly, there was no inhibition of the uptake by anti-CD11b (Mac-1 integrin) antibody. Polyanionic scavenger receptor ligands heparin, polyinosinic acid, fucoidan and dextran sulfate decreased the uptake of SPIO by J774A.1 macrophages and Kupffer cells by 60–75%. These data unambiguously show that SPIO is taken up via interaction by scavenger receptors, but not via dextran recognition by carbohydrate receptors. Understanding of nanoparticle-receptor interaction can provide guidance for the design of long circulating, non-toxic nanomedicines.

Keywords Macrophage • Iron oxide • Scavenger receptor • Dextran • Kupffer cell • Nanoparticle • Lectin

---

*Ying Chao and Priya Prakash Karmali contributed equally to this chapter.

Y. Chao • D. Simberg (✉)
Moores Cancer Center, University of California San Diego, La Jolla, CA, USA
e-mail: dsimberg@ucsd.edu

P.P. Karmali
Cancer Research Center, Sanford-Burnham Medical Research Institute, La Jolla, CA 92037, USA

E. Zahavy et al. (eds.), *Nano-Biotechnology for Biomedical and Diagnostic Research*, Advances in Experimental Medicine and Biology 733, DOI 10.1007/978-94-007-2555-3_11,

11.1 Introduction

Sequestration of nanoscale drug/gene delivery systems by macrophages is one of the most serious limitations of in vivo nanomedicine (Moghimi et al. 2001). Multiple defense mechanisms with overlapping specificities efficiently recognize and sequester the injected foreign materials (Taylor et al. 2005). The macrophage recognition can strongly reduce the effectiveness of the iron oxide nanoparticulate-based imaging and treatments and can result in the damage to the peripheral immune cells (Dobrovolskaia and McNeil 2007; Berry et al. 2004; Oberdorster et al. 2005). Dextran-coated superparamagnetic iron oxide nanoparticles (hereafter SPIO) are widely used in the clinic as MRI contrast agents (e.g., ferumoxides, Feridex I.V.). Dextran-coated SPIO consists of two main chemical components: Fe_3O_4 core crystals of 5–10-nm size embedded in a meshwork of branched dextran (10–40 kDa). Fe_3O_4 crystals are presumably negatively charged due to the ionization of hydrated ferric oxide (Jung 1995; Jung and Jacobs 1995).

Systemically injected SPIO is rapidly cleared from systemic circulation by resident liver and spleen macrophages. SPIO in plasma become enriched by high molecular weight kininogen, histidine-proline rich glycoprotein, beta-2 glycoprotein, and mannose binding lectins (MBL) A and C (Simberg et al. 2009). Histidine-rich proteins are known to bind to anionic surfaces and divalent metal salts (Colman and Schmaier 1997; Schousboe 1985), while MBL is able to recognize both polyanions and sugars (Kuroki et al. 1997; Tan et al. 1996). Some reports suggest that plasma protein corona plays an important role in the nanoparticle recognition by macrophages, as the cells actually "see" the absorbed proteins rather than the nanoparticle itself (Cedervall et al. 2007; Chonn et al. 1992; Moghimi et al. 2006; Lynch et al. 2009). However, other evidence suggests that the macrophage uptake of SPIO is independent of common plasma opsonins and the mechanism likely involves direct binding of SPIO to macrophages (Simberg et al. 2009).

Nevertheless, the chemical component of the SPIO surface that triggers the recognition by macrophages, as well as the receptor(s) responsible for this process are not well understood. There is evidence that macrophage uptake of dextran iron oxide nanoparticles is mediated by scavenger receptors (Raynal et al. 2004). Scavenger receptors have been implicated in the recognition of pathogen-associated molecular patterns, often polyanionic in nature. At the same time, it is not clear whether carbohydrate recognition receptors participate in SPIO uptake, considering the fact that dextran is a bacterial polysaccharide. Pattern recognition receptors, such as Toll-like receptors and C-type lectins, are able to recognize bacterial and fungal polysaccharides, although some of these receptors lack phagocytic function (Le Cabec et al. 2005). SIGN-R1 has been shown to be involved in the uptake of dextrans by spleen macrophages (Kang et al. 2004). It has been reported that SPIO and liposome uptake can be mediated via Mac-1 integrin CD11b/CD18 receptor (von Zur Muhlen et al. 2007), which has a lectin-like activity (Ross 2002).

Here, we set out to understand the role of sugar recognition pathway in SPIO uptake. We systematically studied the effect of carbohydrate receptors on the uptake of the nanoparticles. Understanding the mechanisms of nanoparticle immune recognition could improve the biodistribution and tumor delivery of therapeutic and diagnostic nanoparticles.

11.2 Materials and Methods

11.2.1 Nanoparticles

SPIO nanoparticles coated with dextran T-10 (10-kDa) were obtained from the Department of Radiology UCSD as Feridex I.V. (Ferumoxides) solution. Particles were stored in PBS at 4°C prior to use. Size and ζ (zeta)-potential of nanoparticles before and after dextran modification were measured

Table 11.1 Set of primers used for reverse transcription PCR amplification of receptor mRNA in macrophages and transfected cells

Species	Gene	Accession number	Forward primer	Reverse primer
Mus musculus	MSR1 (SR-A1)	NM_031195	TGA ACG AGA GGA TGC TGA CTG	TGT CAT TGA ACG TGC GTC AAA
	MGL-1	NM_010796	TGA GAA AGG CTT TAA GAA CTG GG	GAC CAC CTG TAG TGA TGT GGG
	Dectin-1	NM_020008	GAC TTC AGC ACT CAA GAC ATC C	TTG TGT CGC CAA AAT GCT AGG
	SIGNR1	NM_133238	CTG GCG TAG ATC GAC TGT GC	AGA CTC CTT GCT CAT GTC AAT G
	GAPDH	NM_008084	AGG TCG GTG TGA ACG GAT TTG	TGT AGA CCA TGT AGT TGA GGT CA
	endo180	NM_008626	ATC CAG GGA AAC TCA CAC GGA	GCG CTC ATC TTT GCC GTA GT

using Zetasizer Nano (Malvern, UK). For nanoparticle imaging with transmission electron microscopy, 5 μl of 0.5 mg/ml nanoparticle solution in water were placed on the Formvar/carbon coated grids (Ted Pella, Redding, CA). After 1 min the grid was gently blotted and air-dried. The samples were studied without counterstaining under the Hitachi 600A electron microscope, at 75 kV and different instrumental magnifications. Images were captured by the 11.2 Megapixel cooled CCD camera (SIA, Atlanta) controlled by the MaxIm DL v. 5.2 software (Diffraction Limited, Ottawa, Ontario, Canada).

11.2.2 Receptor Clones

pCMV-SPORT6 plasmid encoding mouse macrophage galactose N-acetyl-galactosamine specific lectin 1 MGL-1 (BC014811) and mouse C-type lectin msDectin 1 (BC027742) were obtained from Open Biosystems. pCMV-SPORT6 plasmid encoding mouse macrophage scavenger receptor AI (short splice variant B) was obtained from ATCC (MGC-6140). Full-length mouse SIGNR1 cDNA was amplified using the following primers: 5′-GCAGCTAGCATGAGTGATTCTAAGGAAATGGGGAAG-3′ and 5′GCATCTCGAGTCACTTGCTAGGGCAGGAAGTT3′. RNA was isolated from J774A.1 cells using a Total RNA isolation kit (Qiagen). cDNA was synthesized from RNA using Bio-Rad iScript cDNA synthesis kit and amplified using High Fidelity PCR kit (Roche). cDNA was subcloned into pcDNA3.1 vector (Invitrogen) using T4 DNA ligase. The receptor expression was tested in J774A.1 macrophages and Kupffer cells by reverse transcriptase PCR (Table 11.1).

11.2.3 In Vitro Nanoparticle Uptake Experiments

J774A.1 and HEK293T cells were maintained in DMEM/high glucose media (Hyclone) supplemented with 10% fetal bovine serum and 1% penicillin-streptomycin-L-glutamine solution (Invitrogen). Kupffer cells were isolated from collagenase-perfused mouse liver by a differential centrifugation method (Simberg et al. 2009). For receptor expression experiments, HEK 293T cells were seeded at a density of 0.5–1 × 10^6 cells/well in 12 or 6 well plates and transiently transfected with 1.6–4 μg of receptor plasmids or empty vector plasmids (control) using Lipofectamine 2000 (Invitrogen) per the manufacturer's instructions. The mRNA expression of SR-AI, MGL-1, SIGNR1 and msDectin-1 receptors was verified with RT-PCR using the set of primers from Table 11.1. The transfected cells were incubated 24 h later with 0.1 mg/ml (Fe concentration) of nanoparticles for 2 h in complete medium at 37°C. At the end of the incubation, the cells were washed two times with serum free DMEM followed once with PBS, lysed with 10% SDS, and iron oxide uptake was quantified using QuantiChrom Iron Assay (BioAssay Systems). Nanoparticle uptake experiments in J774A.1 cells or in isolated Kupffer cells were performed similarly.

For ligand inhibition experiments, polyinosinic acid, dextran sulfate 500-kDa, heparin, fucoidan, and mannan (all from Sigma) were added to J774A.1 cells 15 min prior to the addition of nanoparticles. For antibody inhibition, the cells were incubated with anti macrophage mannose receptor CD206 (R&D AF2535) or anti CD11b (AbD Serotec, MCA711G) at 20 μg/ml. For trypsin inhibition, the cells were treated with 1 mg/ml trypsin (Sigma) for 10 min and washed two times in serum-free medium prior to the incubation with the nanoparticles. All uptake experiments were performed in complete medium supplemented with 10% fetal calf serum except for the CD11b inhibition which was performed with 20% mouse plasma, 2 mM Ca++ and 10 μM PPAK (thrombin inhibitor).

11.3 Results and Discussion

11.3.1 Characterization of SPIO (Feridex) Nanoparticles

Superparamagnetic iron oxide nanoparticles coated with low molecular weight (10 kDa) dextran (Feridex, Ferumoxides) exhibit strong uptake by cultured macrophages *in vitro* and by the spleen and liver macrophages *in vivo* (Simberg et al. 2009). The particles were filtered through 0.1 μm filter and characterized by transmission electron microscopy (Fig. 11.1). Similar to the previous study (Jung 1995; Jung and Jacobs 1995), the particles appeared as clusters of 50–100 nm aggregates of electron-dense crystalline Fe_3O_4 lattices embedded in a meshwork of dextran (not visible on a TEM image). The hydrodynamic size of the particles was 85 nm (polydispersity index 0.241), and ζ (zeta)-potential was −13.1 mV.

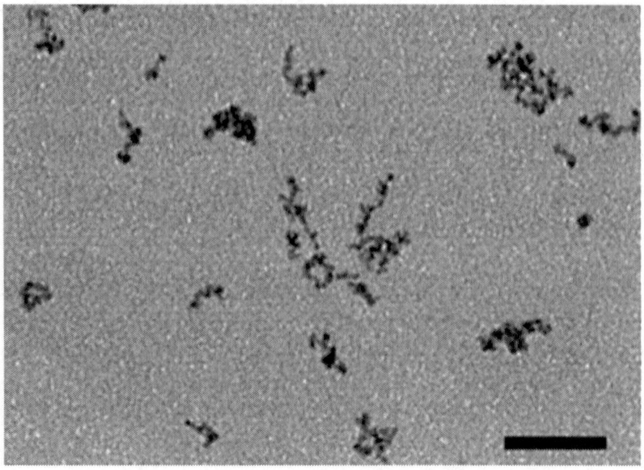

Fig. 11.1 Transmission electron microscopy of Feridex particles. The sample was prepared as described in Methods. The particles appear as heterogeneous clusters of electron-dense iron oxide crystals. Size bar 100 nm

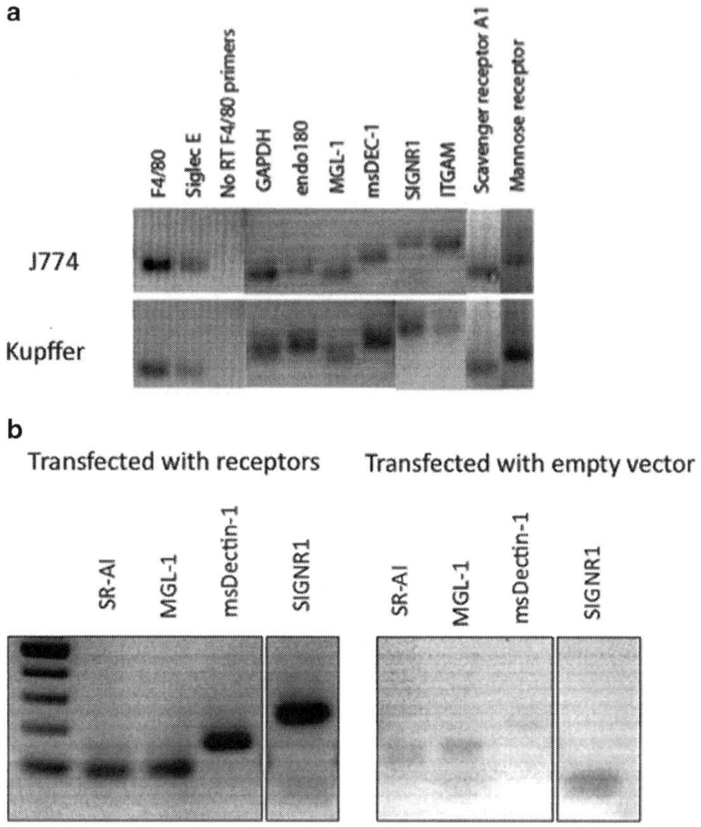

Fig. 11.2 Expression of receptors in macrophages and transfected cells Reverse transcription PCR was performed as described in Methods. (**a**) Intrinsic expression of receptors and other markers in J774A.1 and Kupffer cells. F4/80 is a common macrophage marker, and GADPH was used as a housekeeping gene. (**b**) Expression following transient transfection of the receptors into HEK293T cells. All transfected receptors showed expression in the host cells

11.3.2 Effect of Expression of Innate Immunity Receptors on SPIO Uptake

In order to understand the role of innate immunity receptors in SPIO uptake, we first tested the intrinsic macrophage expression of SR-A1, MGL-1, msDectin, SIGNR1 and macrophage mannose receptor (MMR) in J774A.1 macrophages and in Kupffer cells. According to the RT-PCR results (Fig. 11.2a), the mRNA of the receptors was present in both cell types. To understand the role of these receptors in the uptake of SPIO, we transiently transfected HEK293T cells with the plasmids coding for the above receptors using lipofection. MMR plasmid is not commercially available and we have not been able to clone MMR cDNA into a vector, possibly due to the large size of the receptor. Twenty-four hours following transfection, we tested the mRNA expression by RT-PCR (Fig. 11.2b).

Since SPIO is coated with the carbohydrate dextran, we tested the ability of carbohydrate receptors MGL-1, SIGN-R1 and msDectin-1 to recognize and promote the uptake of SPIO. Transient expression of macrophage receptors in non-macrophage cells is a well-established method to demonstrate function, including uptake of nanoparticles (Jozefowski et al. 2005). The transfected cells were incubated with 0.1 mg/ml SPIO and the level of uptake was quantified by iron assay. Transient transfection of these receptors in HEK293T cells did not promote the uptake of SPIO (Fig. 11.3a). As shown in

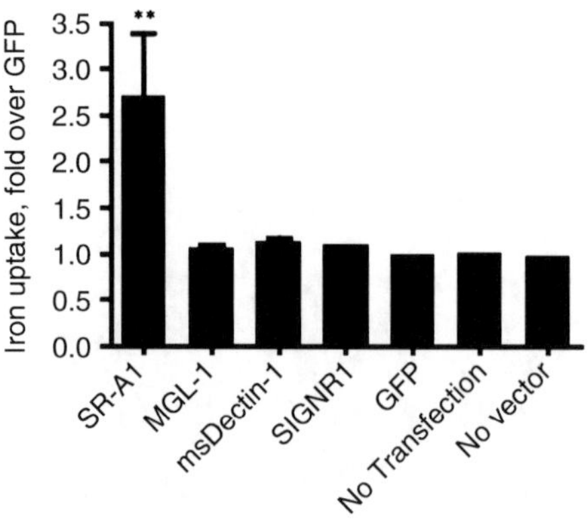

Fig. 11.3 Effect of transiently expressed scavenger receptors on SPIO binding and uptake HEK293T cells were transiently transfected with receptor cDNAs and then incubated with 0.1 mg/ml SPIO (Feridex) as described in Methods. The uptake of SPIO by cells transfected with macrophage receptors was quantified with QuantiChrom™ Iron Assay Kit (see Methods). Control was GFP-transfected cells; P value = 0.0082, t-test, n = 4

Fig. 11.3, SR-A1 expression resulted in almost three-fold increase in the binding and uptake of SPIO compared to non-transfected cells and GFP-transfected cells (P value = 0.0082).

Many types of negatively charged nanoparticulate ligands, including gold, polystyrene and silica nanoparticles appear to be recognized via the scavenger receptor mechanism (Kanno et al. 2007; Nagayama et al. 2007; Raynal et al. 2004; Demoy et al. 1999). Scavenger receptor SR-AI/II recognizes negatively charged surfaces through positively charged collagen-like domain (CLD) (Doi et al. 1993; Kanno et al. 2007; Nagayama et al. 2007; Raynal et al. 2004), and this domain could participate in SPIO uptake.

11.3.3 Effect of Ligand and Blocking Antibody Inhibition of Uptake of SPIO

Next, we used J774A.1 murine monocyte cell line for the uptake studies. The role of carbohydrate receptor pathway (mainly mannose receptor CD206) was tested with ligands and blocking antibodies. We tested the uptake of SPIO in the presence of 20-kDa dextran, mannan (CD206 ligand), and anti-CD206 blocking antibody. In all cases, there was no effect on the SPIO uptake (Fig. 11.4). The same result was obtained with isolated murine Kupffer cells.

In order to test the previously suggested role of Mac-1 integrin (CD11b/CD18) in the uptake of SPIO (von Zur Muhlen et al. 2007), we used anti-CD11b blocking antibody to inhibit the uptake. The uptake was unmodified whether the experiment was performed in normal cell medium or in mouse plasma (Fig. 11.4). In the previous study, activated macrophages from atherosclerotic lesions were used. Mac-1 has a lectin-like domain in its the CD11b subunit that mediates adhesion between macrophages and other cells (Ross 2002), and it has been shown to play a role in newly recruited (activated) macrophages and not in resident Kupffer cells (Moghimi 2002; von Zur Muhlen et al. 2007).

In contrast with carbohydrate receptor inhibitors, addition of polyanionic scavenger receptor ligands fucoidan (10 μg/ml), poly-inosinic acid (10 μg/ml), heparin (10 U/ml) or dextran sulfate (3 μg/ml) to the cells prior to SPIO produced between 65% and 76% inhibition (P value = 0.0003 for dextran sulfate) of uptake in J774A.1 cells (Fig. 11.5). Similar levels of inhibition by polyanions were observed for Kupffer cells (not shown). Pretreatment of J774A.1 cells with 0.1 mg/ml trypsin reduced

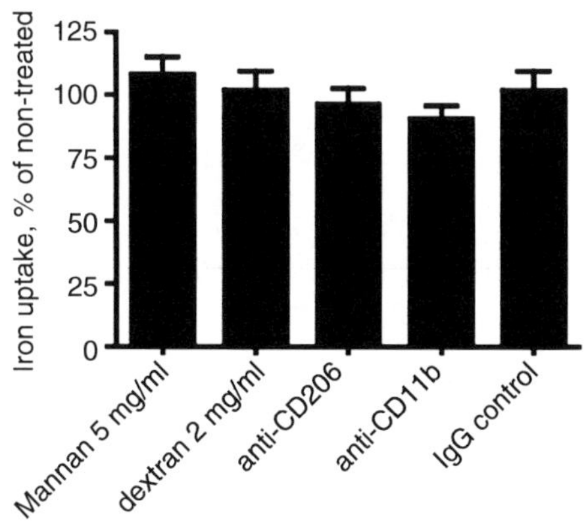

Fig. 11.4 Effect of carbohydrate receptor inhibitors on SPIO uptake by J774A.1 cells The experiment was performed as described in Methods. CD11b antibody was tested in plasma protein-rich media. Inhibitors of mannose receptor (CD206) or Mac-1 integrin (CD11b/CD18) did not reduce the uptake of SPIO. Similar result was obtained on Kupffer cells (not shown)

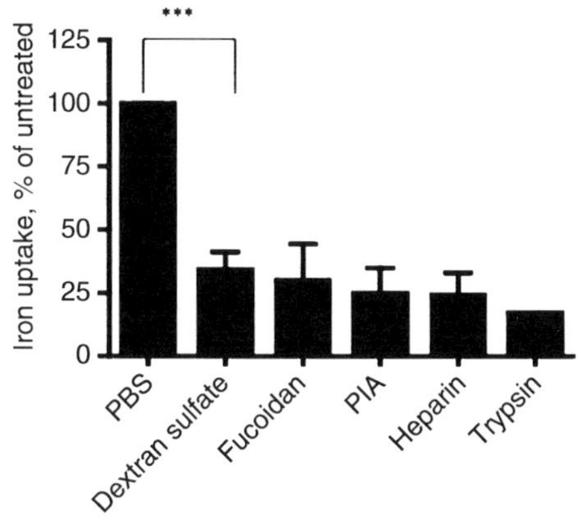

Fig. 11.5 Effect of scavenger receptor inhibitors on SPIO uptake by J774A.1 macrophages Cells were incubated in complete medium for 1–2 h with the corresponding ligands (for concentrations, see Results), or with 0.1 mg/ml trypsin followed by 0.1 mg/ml SPIO (Feridex), then washed and the iron uptake was quantified. A SPIO uptake by the macrophages was significantly reduced in the presence of polyanionic scavenger receptor ligands (for dextran sulfate P value=0.0003, t-test, n=3), suggesting charge-dependent uptake. The trypsin effect demonstrates that the uptake is protein (receptor)-dependent

the uptake by about 80% (Fig. 11.5). These data suggest that the mechanism of SPIO uptake is scavenger receptor-dependent.

11.4 Conclusions

Here, we set out to investigate the role of carbohydrate recognition in the uptake of dextran coated superparamagnetic iron oxide. Taken together, our data suggest that the uptake of SPIO nanoparticles does not involve recognition by lectin-like receptors. The expression of lectin-like receptors and use of soluble inhibitors and even dextran itself did not affect the SPIO uptake. We also did not find evidence that SPIO uptake is mediated via Mac-1 (CR3 or CD11b/CD18) receptor as was suggested previously (von Zur Muhlen et al. 2007). These data provide insight on the mechanisms of nanoparticle immune recognition and clearance in vitro and in vivo.

Acknowledgements Authors wish to thank Dr. Erkki Ruoslahti from Sanford-Burnham Medical Research Institute at UCSB for his relentless support and guidance throughout the project. Dr. Yuko Kono from UCSD Department of Radiology is acknowledged for providing Feridex I.V. The work was supported by NIH NCI grants CA119335 and CA124427.

References

Berry, C. C., Wells, S., Charles, S., Aitchison, G., & Curtis, A. S. (2004). Cell response to dextran-derivatised iron oxide nanoparticles post internalisation. *Biomaterials, 25*, 5405–5413.

Cedervall, T., Lynch, I., Lindman, S., Berggard, T., Thulin, E., Nilsson, H., Dawson, K. A., & Linse, S. (2007). Understanding the nanoparticle-protein corona using methods to quantify exchange rates and affinities of proteins for nanoparticles. *Proceedings of the National Academy of Sciences of the United States of America, 104*, 2050–2055.

Chonn, A., Semple, S. C., & Cullis, P. R. (1992). Association of blood proteins with large unilamellar liposomes in vivo. Relation to circulation lifetimes. *The Journal of Biological Chemistry, 267*, 18759–18765.

Colman, R. W., & Schmaier, A. H. (1997). Contact system: A vascular biology modulator with anticoagulant, profibrinolytic, antiadhesive, and proinflammatory attributes. *Blood, 90*, 3819–3843.

Demoy, M., Andreux, J. P., Weingarten, C., Gouritin, B., Guilloux, V., & Couvreur, P. (1999). In vitro evaluation of nanoparticles spleen capture. *Life Sciences, 64*, 1329–1337.

Dobrovolskaia, M. A., & McNeil, S. E. (2007). Immunological properties of engineered nanomaterials. *Nature Nanotechnology, 2*, 469–478.

Doi, T., Higashino, K., Kurihara, Y., Wada, Y., Miyazaki, T., Nakamura, H., Uesugi, S., Imanishi, T., Kawabe, Y., Itakura, H., et al. (1993). Charged collagen structure mediates the recognition of negatively charged macromolecules by macrophage scavenger receptors. *The Journal of Biological Chemistry, 268*, 2126–2133.

Jozefowski, S., Arredouani, M., Sulahian, T., & Kobzik, L. (2005). Disparate regulation and function of the class a scavenger receptors SR-AI/II and MARCO. *Journal of Immunology, 175*, 8032–8041.

Jung, C. W. (1995). Surface properties of superparamagnetic iron oxide MR contrast agents: Ferumoxides, ferumoxtran, ferumoxsil. *Magnetic Resonance Imaging, 13*, 675–691.

Jung, C. W., & Jacobs, P. (1995). Physical and chemical properties of superparamagnetic iron oxide MR contrast agents: Ferumoxides, ferumoxtran, ferumoxsil. *Magnetic Resonance Imaging, 13*, 661–674.

Kang, Y. S., Kim, J. Y., Bruening, S. A., Pack, M., Charalambous, A., Pritsker, A., Moran, T. M., Loeffler, J. M., Steinman, R. M., & Park, C. G. (2004). The C-type lectin SIGN-R1 mediates uptake of the capsular polysaccharide of streptococcus pneumoniae in the marginal zone of mouse spleen. *Proceedings of the National Academy of Sciences of the United States of America, 101*, 215–220.

Kanno, S., Furuyama, A., & Hirano, S. (2007). A murine scavenger receptor MARCO recognizes polystyrene nanoparticles. *Toxicological Sciences, 97*, 398–406.

Kuroki, Y., Honma, T., Chiba, H., Sano, H., Saitoh, M., Ogasawara, Y., Sohma, H., & Akino, T. (1997). A novel type of binding specificity to phospholipids for rat mannose-binding proteins isolated from serum and liver. *FEBS Letters, 414*, 387–392.

le Cabec, V., Emorine, L. J., Toesca, I., Cougoule, C., & Maridonneau-Parini, I. (2005). The human macrophage mannose receptor is not a professional phagocytic receptor. *Journal of Leukocyte Biology, 77*, 934–943.

Lynch, I., Salvati, A., & Dawson, K. A. (2009). Protein-nanoparticle interactions: What does the cell see? *Nature Nanotechnology, 4*, 546–547.

Moghimi, S. M. (2002). Liposome recognition by resident and newly recruited murine liver macrophages. *Journal of Liposome Research, 12*, 67–70.

Moghimi, S. M., Hunter, A. C., & Murray, J. C. (2001). Long-circulating and target-specific nanoparticles: Theory to practice. *Pharmacological Reviews, 53*, 283–318.

Moghimi, S. M., Hamad, I., Andresen, T. L., Jorgensen, K., & Szebeni, J. (2006). Methylation of the phosphate oxygen moiety of phospholipid-methoxy(polyethylene glycol) conjugate prevents PEGylated liposome-mediated complement activation and anaphylatoxin production. *The FASEB Journal, 20*, 2591–2593.

Nagayama, S., Ogawara, K., Minato, K., Fukuoka, Y., Takakura, Y., Hashida, M., Higaki, K., & Kimura, T. (2007). Fetuin mediates hepatic uptake of negatively charged nanoparticles via scavenger receptor. *International Journal of Pharmaceutics, 329*, 192–198.

Oberdorster, G., Oberdorster, E., & Oberdorster, J. (2005). Nanotoxicology: An emerging discipline evolving from studies of ultrafine particles. *Environmental Health Perspectives, 113*, 823–839.

Raynal, I., Prigent, P., Peyramaure, S., Najid, A., Rebuzzi, C., & Corot, C. (2004). Macrophage endocytosis of superparamagnetic iron oxide nanoparticles: Mechanisms and comparison of ferumoxides and ferumoxtran-10. *Investigative Radiology, 39*, 56–63.

Ross, G. D. (2002). Role of the lectin domain of Mac-1/CR3 (CD11b/CD18) in regulating intercellular adhesion. *Immunologic Research, 25*, 219–227.

Schousboe, I. (1985). Beta 2-Glycoprotein I: A plasma inhibitor of the contact activation of the intrinsic blood coagulation pathway. *Blood, 66*, 1086–1091.

Simberg, D., Park, J. H., Karmali, P. P., Zhang, W. M., Merkulov, S., McCrae, K., Bhatia, S. N., Sailor, M., & Ruoslahti, E. (2009). Differential proteomics analysis of the surface heterogeneity of dextran iron oxide nanoparticles and the implications for their in vivo clearance. *Biomaterials, 30*, 3926–3933.

Tan, S. M., Chung, M. C., Kon, O. L., Thiel, S., Lee, S. H., & Lu, J. (1996). Improvements on the purification of mannan-binding lectin and demonstration of its Ca(2+)-independent association with a C1s-like serine protease. *Biochemical Journal, 319*(Pt 2), 329–332.

Taylor, P. R., Martinez-Pomares, L., Stacey, M., Lin, H. H., Brown, G. D., & Gordon, S. (2005). Macrophage receptors and immune recognition. *Annual Review of Immunology, 23*, 901–944.

Von Zur Muhlen, C., Von Elverfeldt, D., Bassler, N., Neudorfer, I., Steitz, B., Petri-Fink, A., Hofmann, H., Bode, C., & Peter, K. (2007). Superparamagnetic iron oxide binding and uptake as imaged by magnetic resonance is mediated by the integrin receptor Mac-1 (CD11b/CD18): implications on imaging of atherosclerotic plaques. *Atherosclerosis, 193*, 102–11.

Chapter 12
Toxicity of Gold Nanoparticles on Somatic and Reproductive Cells

U. Taylor, A. Barchanski, W. Garrels, S. Klein, W. Kues, S. Barcikowski, and D. Rath

Abstract Along with the number of potential applications for gold nanoparticles (AuNP) especially for medical and scientific purposes, the interest in possible toxic effects of such particles is rising. The general perception views nanosized gold colloids as relatively inert towards biological systems. However, a closer analysis of pertinent studies reveals a more complex picture. While the chemical compound of which the nanoparticles consists plays an important role, further biocompatibility determining aspects have been made out. The vast majority of trials concerning AuNP-toxicity were performed using somatic cell culture lines. The results show a considerable dependency of toxic effects on size, zeta potential and surface functionalisation. In vivo studies on this subject are still rare. Based on the existing data it can be assumed, that a dosage of under <400 µg Au/kg showed no untoward effects. If higher amounts were applied toxicity depended on route of administration and particle size. Since nanoparticles have been shown to cross reproduction-relevant biological barriers such as the blood-testicle and the placental barrier the question of their reprotoxicity arises. Yet data concerning this subject is far from adequate. Regarding gametes, recent experiments showed a dose-dependent sensitivity of spermatozoa towards AuNP. Oocytes have not yet been tested in that respect. Interestingly, so far no effects were detected on embryos after gold nanoparticle exposure. In conclusion, the biocompatibility of gold nanoparticles depends on a range of particle specific aspects as well as the choice of target tissue. Further clarification of such matters are subject to ongoing research.

Keywords Gold nanoparticles • Toxicity • Reprotoxicity • Ligand-free • Pulsed laser ablation in liquids • Cell culture • Mouse • Gametes • Embryos

12.1 Introduction

Gold nanoparticles have triggered an emerging interest for medical and scientific purposes because of their outstanding characteristics due to their electronic, optical, magnetic and catalytic properties when compared with corresponding bulk material. Among the most popular application areas within

U. Taylor (✉) • W. Garrels • S. Klein • W. Kues • D. Rath
Institute of Farm Animal Genetics, Friedrich-Loeffler-Institut, Höltystrasse 10, 31535 Neustadt-Mariensee, Germany
e-mail: ustendel@gmx.de

A. Barchanski • S. Barcikowski
Laserzentrum Hannover e.V., Hollerithallee 8, 30419 Hannover, Germany

E. Zahavy et al. (eds.), *Nano-Biotechnology for Biomedical and Diagnostic Research*, Advances in Experimental Medicine and Biology 733, DOI 10.1007/978-94-007-2555-3_12,

life sciences are selective coupling (Sokolov et al. 2003) and sensing (Wang and Ma 2009) of target molecules, localized cancer therapy by plasmonic heating of malignant tissue (Gannon et al. 2008) and delivery of effector molecules to specific receptors or areas of interest (Han et al. 2007). However, since many potential applications for gold nanoparticles are likely to be performed on living cells or organisms, the question of their biocompatibility is of high relevance.

Unintended effects of nanoscaled particles seem to derive mainly from their higher mass-specific surface area, which renders them more biologically active than larger particles of the same chemistry, with a surface-specific dose–response (Faux et al. 2003; Oberdörster et al. 2005.). The underlying mechanisms for nanoparticle-related cellular damage suggested in recent literature are the production of reactive oxygen species (ROS) (Oberdörster et al. 2005) and interaction with DNA (Singh et al. 2009). The interactions at the nano-bio interface, which ultimately determine the toxic potential of any given nanoparticle, are driven by a multitude of parameters (Nel et al. 2009). Some of them are predestined by the nanoparticle itself, such as chemical composition, surface functionalization, size, shape and polarity. Others are influenced by the suspending medium, which most importantly applies to the surface charge. Therefore, it is little surprising that the results of gold nanoparticle biocompatibility studies display a certain amount of heterogeneity, since the used particles, even though they were all based on gold nanoparticles, differed in many aspects, like size and surface functionalization. The influence of AuNP on cell cultures has been examined most extensively and therefore shows exemplarily the diverse effect such particles can potentially have. However, while in somatic cells insults derived from nanoparticle exposure may cause inflammation or even malignant transformation, in case of germline cells, either defect might lead to impaired fertility and/or congenital defects in the offspring. Thus, the current knowledge about reproductive nanotoxicology shall also be summarized.

12.2 Effect of Gold Nanoparticles on Somatic Cells

12.2.1 Cell Culture Studies

Looking at the results of different cell culture studies, the considerable influence of the above mentioned individual particle properties on the harmful potential of gold nanoparticles could be sensed. In most of the trials the effect of differently composed or sized gold nanoparticles were compared with each other. The observed toxicity ranged from negligible, regardless of the used particle type (Shenoy et al. 2006; Salmaso et al. 2009) to intermediate (Thomas and Klibanov 2003; Connor et al. 2005; Massich et al. 2010; Taylor et al. 2010b) and even severe (Pan et al. 2007; Patra et al. 2007; Ding et al. 2010) (Table 12.1). The study performed by Pan et al. (2007) is a distinct example how the effect of nanoparticle on cells can be driven by size, showing that even a moderate decrease in size (from 1.8 nm to 1.4 nm) can make particles four to six times more noxious (Pan et al. 2007). Trials conducted by Ding et al. (2010) exemplify nicely the impact of the nanoparticles zeta potential, i.e. its electric potential at the particle-liquid-interface, on the outcome of the study, elucidating that cytotoxicity correlated with an increasing in positive charge. The experiments from Massich et al. (2010) indicate the influence of surface functionalization, detecting a cytotoxic effect of gold nanoparticle in conjunction with citrate, a common stabilizing agent in chemically derived nanoparticles, which is quite remarkable, because the employed gold nanoparticle dosage was with 10 nM gold fairly low. Only in two studies nanoparticles entirely without any surface modification were used. In these cases the nanoparticles were produced by pulsed laser ablation in liquids (PLAL) without the need for any stabilizing agent. Salmaso et al. (2009) did not observe any toxicity with such particles up to a concentration of 0.74 nM gold. However, Taylor et al. (2010b) noticed a cytotoxic effect, but only at concentrations five magnitudes higher than the ones tested by Salmaso et al. (2009) (Fig. 12.1).

Table 12.1 Overview of several toxicological studies run in various cell lines concerning gold nanoparticles

Reference	Cell line	Surface modification	NP concentration and size	Exposure duration	Tests	Results
Thomas and Klibanov 2003	COS-7 cells	PEI2	N/A N/A	6 h, 42 h	MTT	20–30% loss of viability
Tkachenko et al. 2003	Human liver carcinoma, HepG2	BSA, 4 targeting peptides	N/A d = 20–25 nm	12 h	LDH	5% loss of viability
Connor et al. 2005	Leukemia cell line, K562	Citrate, Biotin, L-cysteine, Glucose, CTAB	0–250 μM Au d = 4, 12, 18 nm	3 days	MTT	Toxic, only if modified with glucose and cystein and NP-concentration > 25 μM
Fu et al. 2005	Human breast carcinoma xenograft cells, MDA-MB-231	Coumarin-PEG-thiol	50–200 μg/ml d = 10 nm	24 h	Cell Titer 96	No toxicity detected
Shenoy et al. 2006	Human breast carcinoma xenograft cells, MDA-MB-231	Coumarin-PEG-thiol	50–200 μg/ml d = 10 nm	24 h	Cell Titer 96	No toxicity detected
Pan et al. 2007	Cervix carcinoma, HeLa; Melanoma, SK-Mel-28; Mouse fibroblasts, L929; Mouse macrophages (J774A1)	Triphenylphosphine derivates	0–6,300 μM d = 8.0–15 nm	48 h	Microscopy, MTT, Annexin assay	Highest toxicity (three–fold higher than any other size) at 1.4 nm diameter
Patra et al. 2007	Human lung carcinoma, A549; Human liver carcinoma, HepG2; Syrian hamster kidney fibroblasts, BHK21	Citrate	0–120 nM d_{hyd} = 33 nm	48 h	Microscopy, PI, MTT, cleavage of poly (ADP-ribose) polymerase	Toxicity detected in A549 cells only, after exposure to 10 nM

(continued)

Table 12.1 (continued)

Reference	Cell line	Surface modification	NP concentration and size	Exposure duration	Tests	Results
Massich et al. 2010	Cervix carcinoma, HeLa	Citrate, BSA, ssDNA, dsDNA, dsRNA	10 nM d = 15 nm	24 h	Gene expression analysis, cell-cycle analysis, annexin assay	Citrate stabilised nanoparticles caused change in gene expression, disturbance of mitosis and ca. 20% increase in apotosis
Taylor et al. 2010	Bovine endothelial cells, GM7373	none	0–50 µM d = 15 nm	96 h	Microscopy, PI, TUNEL-assay, XTT	Toxicity detected only in XTT-assay after exposure to 50 µM
Ding et al. 2010	Human gastric carcinoma cell line, BGC 823	Chitosan	0.05–1 mg/ml	44 h	MTT	Toxicity detected after exposure to 0.8 mg/ml AuNP with a Zeta-potential of 40 mV, no toxicity with a Zetapotential of 20 and 30 mV

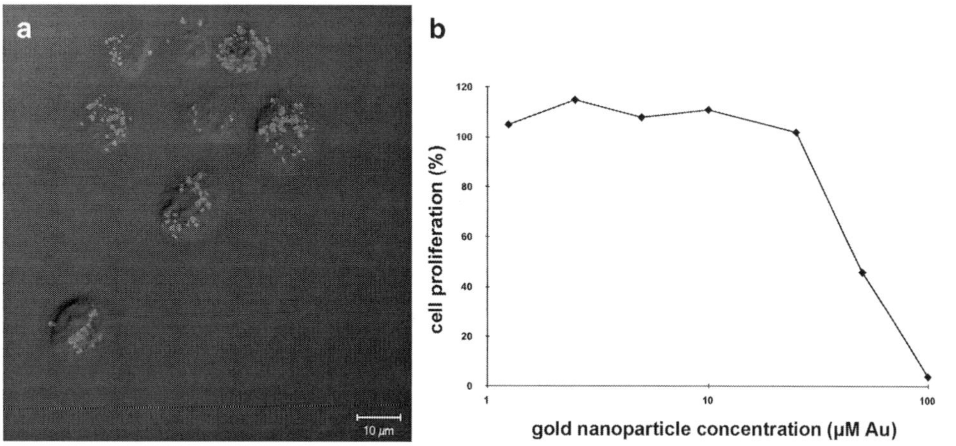

Fig. 12.1 (**a**) Representative laser scanning microscope images of bovine endothelial cells (GM7373) (3D – projections of 10 optical sections (1 µm each)) after 48 h co-incubation with AuNPs (50 µM Au). (**b**) Diagram displaying the results of the XTT Proliferation Assay with bovine endothelial cells (GM7373) as a percentage of living cells (negative control = 100%) against the AuNP concentration on a logarithmic scale (Taylor et al. 2010b)

12.2.2 In-Vivo Response to Gold Nanoparticles

In-vivo toxicity studies with gold nanoparticles are not as numerous yet. All trials were performed with chemically produced particles. But despite the relative similarity of the particles, the results still give very different indications. Low doses (<400 µg/kg) in general seem to cause no appreciable toxicity in mice (Lasagna-Reeves et al. 2010; Zhang et al. 2010). At higher concentrations, reports start to differ, which most probably is due to variations in the experimental set up regarding animal, material and frequency as well as route of administration. The latter is highlighted in the study by Zhang et al. (2010), where the highest level of toxicity was detected if the particles were given orally or injected into the peritoneum, while intravenous injection of gold nanoparticles proved to elicit the least damage (Zhang et al. 2010). Worth noticing is also a mouse study performed by Chen et al. (2009), which observed a drastic effect of gold nanoparticles after intraperitoneal injection, only depending on particle size (Chen et al. 2009). Particles with a diameter of 3, 5, 50 and 100 nm did not show any harmful effects, while sizes between 8 and 37 nm induced severe sickness and shortend the survival time on average to 21 days. This finding once again implies the impact of particle properties other the chemical composition on the potential of nanoparticles to have detrimental effects. No in-vivo trials were run so far comparing the influence other aspects such as surface charge. However, Cho et al. (2009) tested PEG-coated gold nanoparticles and found that they induced acute liver inflammation after a single intravenous injection of 850 µg/kg (Cho et al. 2009).

12.3 Reproductive Toxicology of Gold Nanoparticles

12.3.1 Effect of Gold Nanoparticles on Gametes

So far, there have only been two studies published concerning the impact of gold nanoparticles on gametes. However, both trials concentrated on the male side, i.e. the effect of AuNP on spermatozoa. In each study, one working with chemically derived gold nanoparticles (Wiwanitkit et al. 2009), the other using laser-generated ligand-free particles (Taylor et al. 2010c) a decrease in sperm motility was observed after gold nanoparticle exposure. In the former study, which also reported severe morphological defects in treated spermatozoa, no information has been provided concerning the particle concentration. However, in the latter the particle concentration needed to actually show an effect was 50 μM, which is so high, that it exceeds by far the amount of gold nanoparticles necessary for scientific or medical applications. Moreover, in this case the decrease in motility was not accompanied by an increase in abnormal sperm morphology or impaired membrane integrity (Fig. 12.2). A possible explanation for the apparently more severe toxicity of chemically derived nanoparticles could be that the observed effect is actually due to remnants of the reducing or stabilizing agents used during production, not the nanoparticles themselves. Additionally, current trials indicate an influence of ligand-free AuNP also on the fertilising capability of spermatozoa (Taylor et al., unpublished).

Up to date there are no studies available concerning the impact of gold nanoparticles on oocytes.

12.3.2 Translocation of Gold Nanoparticles to Reproduction-Relevant Sites

Not many studies concerning AuNP biodistribution have examined their ability to pass through reproduction-relevant physiological barriers. So far no information can be given about their ability to enter ovarian follicles. Concerning the crossing of the blood-testicle barrier Balasubramanian et al. (2010) reported the nanoparticle accumulation of AuNP 1 month in the testis thus showing that nanoparticles can potentially cross the blood-testis barrier (Balasubramanian et al. 2010). Translocation of AuNP across the placenta, which acts as another major barrier in reproduction, has been examined after intravenous injection in rodent models. Two of these studies could indeed confirm nanoparticle transfer through the placental membranes (Takahashi and Matsuoka 1981; Semmler-Behnke et al. 2007). Interestingly, another two studies could not find any gold nanoparticles passing through the placental barrier (Challier et al. 1973; Sadauskas et al. 2007), neither did an author who used human placenta in an ex vivo model to investigate the transplacental trafficking of AuNP (Myllynen et al. 2008). Due to the variations in study outcome it is difficult to draw any final conclusions concerning this subject. Therefore, more research is needed to clarify this important matter.

12.3.3 Gold Nanoparticle Impact on Embryo Development

Another crucial aspect is the developmental toxicity of gold nanoparticles. This subject has been addressed in studies using zebrafish (Bar-Ilan et al. 2009; Browning et al. 2009) and chicken (Zielinska et al. 2009, Sawosz et al. 2010) embryos in conjunction with chemically derived particles as well as murine embryos (Taylor et al. 2010a) employing laser-generated nanoparticles. No detrimental effects were noted, even though the presence of AuNP inside the embryos was proven (Bar-Ilan et al. 2009; Browning et al. 2009; Taylor et al. 2010a). However, the data currently available is by far not adequate to properly assess the influence of gold nanoparticles on embryo development because the variety of particle compositions tested is still way to narrow. Furthermore, as yet no long-term in vivo studies have been performed investigating the effect of AuNP on the developing embryo or fetus, respectively. Especially regarding the fact that particle translocation through the placenta cannot be ruled out, there should be an ongoing effort to thoroughly deduce the effect of gold nanoparticles on such a vulnerable organism.

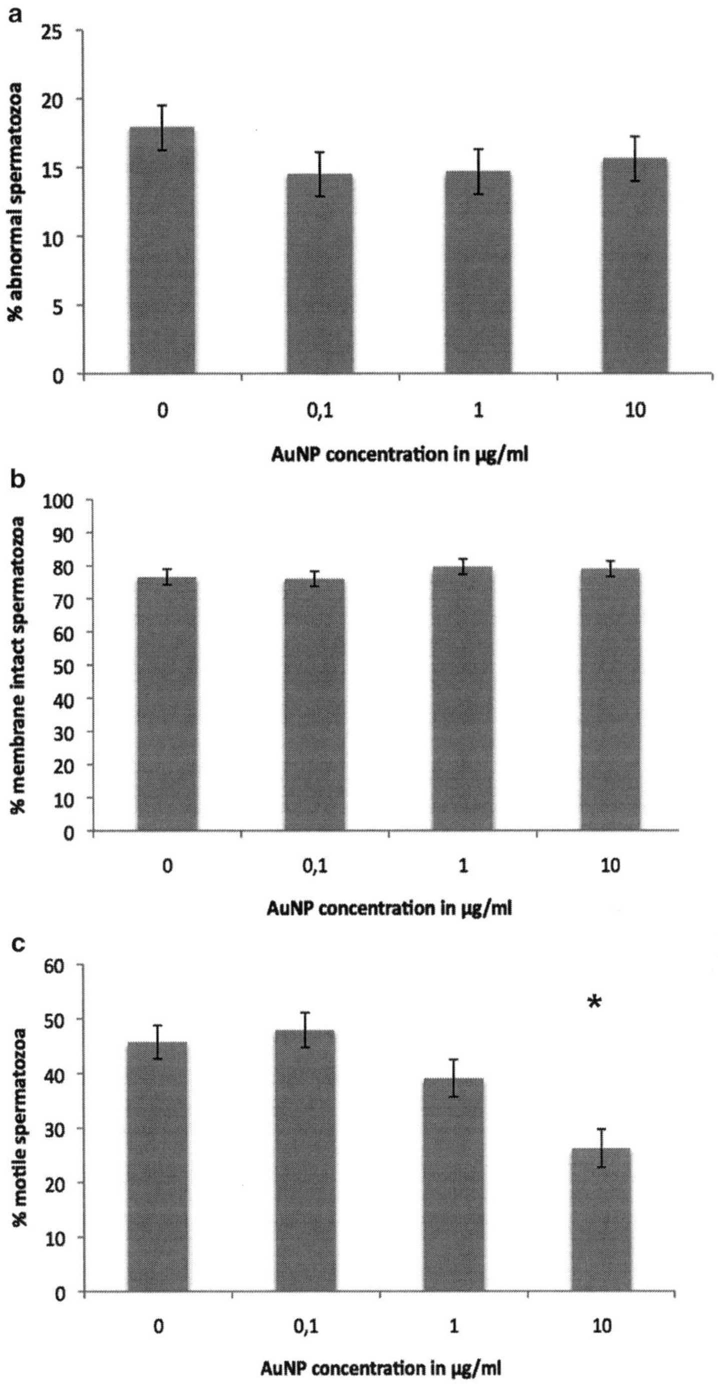

Fig. 12.2 (**a**) Percentage of morphologically abnormal spermatozoa after co-incubation with increasing AuNP concentrations. (**b**) Percentage of membrane intact spermatozoa after co-incubation with increasing AuNP concentrations. (**c**) Percentage of motile spermatozoa after co-incubation with increasing AuNP concentrations (*$p<0.05$) (Taylor et al. 2010c)

12.4 Conclusion

From the provided data it is rather difficult to depict clear trends regarding the biocompatibility of gold nanoparticles. Compared to other metal nanoparticles such as silver (Bar-Ilan et al. 2009), the general toxicity is certainly rather low. However, most studies did observe adverse effects from a certain dosage onwards. Unfortunately, it is rather difficult to compare between studies especially because the information given concerning the dosage are very diverse. It would be recommendable to find a common notion on how to express nanoparticle dosage. One option would be to calculate the particle surface exposed to the cells or the organism as suggested by Oberdörster et al. (2005), since it combines particle number and size and has shown to fit very well in dose–response curves.

References

Balasubramanian, S. K., Jittiwat, J., Manikandan, J., Ong, C. N., Yu, L. E., & Ong, W. Y. (2010). Biodistribution of gold nanoparticles and gene expression changes in the liver and spleen after intravenous administration in rats. *Biomaterials, 31*, 2034–2042.

Bar-Ilan, O., Albrecht, R. M., Fako, V. E., & Furgeson, D. Y. (2009). Toxicity assessments of multisized gold and silver nanoparticles in zebrafish embryos. *Small, 5*(16), 1897–1910.

Browning, L. M., Lee, K. J., Huang, T., Nallathamby, P. D., Lowman, J. E., & Xu, X.-H. N. (2009). Random walk of single gold nanoparticles in zebrafish embryos leading to stochastic toxic effects on embryonic developments. *Nanoscale, 1*, 138–152.

Challier, J. C., Panigel, M., & Meyer, E. (1973). Uptake of colloidal 198Au by fetal liver in rat, after direct intrafetal administration. *International Journal of Nuclear Medicine and Biology, 1*, 103–106.

Chen, Y. S., Hung, Y. C., Liau, I., & Huang, G. S. (2009). Assessment of the in vivo toxicity of gold nanoparticles. *Nanoscale Research Letters, 4*, 858–864.

Cho, W. S., Cho, M., Jeong, J., Choi, M., Cho, H. Y., Han, B. S., Kim, S. H., Kim, H. O., Lim, Y. T., & Chung, B. H. (2009). Acute toxicity and pharmacokinetics of 13 nm-sized PEG-coated gold nanoparticles. *Toxicology and Applied Pharmacology, 236*, 16–24.

Connor, E. E., Mwamuka, J., Gole, A., Murphy, C. J., & Wyatt, M. D. (2005). Gold nanoparticles are taken up by human cells but do not cause acute cytotoxicity. *Small, 1*, 325–327.

Ding, Y., Bian, X. C., Yao, W., Li, R. T., Ding, D., Hu, Y., Jiang, X. Q., & Hu, Y. Q. (2010). Surface-potential-regulated transmembrane and cytotoxicity of chitosan/gold hybrid nanospheres. *ACS Applied Materials & Interfaces, 2*, 1456–1465.

Faux, S. P., Tran, C. L., Miller, B. G., Jones, A. D., Montellier, C., & Donaldson, K. (2003). *In vitro determinants of particulate toxicity: The dosemetric for poorly soluble dusts*. Sudbury/Suffolk: HSE Books.

Fu W., Shenoy D., Li J., Crasto C., Jones G., DiMarzio C., Sridhar S., Amiji M. (2005). Biomedical applications of gold nanoparticles functionalized using hetero-bifunctional poly (ethylene glycol) spacer. *Proceedings of Material Research Symposium*, 845.

Gannon, C. J., Patra, C. R., Bhattacharya, R., Mukherjee, P., & Curley, S. A. (2008). Intracellular gold nanoparticles enhance non-invasive radiofrequency thermal destruction of human gastrointestinal cancer cells. *Journal of Nanobiotechnology, 6*, 2.

Han, G., Ghosh, P., & Rotello, V. M. (2007). Functionalized gold nanoparticles for drug delivery. *Nanomedicine-UK, 2*, 113–123.

Lasagna-Reeves, C., Gonzalez-Romero, D., Barria, M. A., Olmedo, I., Clos, A., Ramanujam, V. M. S., Urayama, A., Vergara, L., Kogan, M. J., & Soto, C. (2010). Bioaccumulation and toxicity of gold nanoparticles after repeated administration in mice. *Biochemical Biophysical Research Communications, 393*, 649–655.

Massich, M. D., Giljohann, D. A., Schmucker, A. L., Patel, P. C., & Mirkin, C. A. (2010). Cellular response of polyvalent oligonucleotide-gold nanoparticle conjugates. *ACS Nano, 4*, 5641–5646.

Myllynen, P. K., Loughran, M. J., Howard, C. V., Sormunen, R., Walsh, A. A., & Vahakangas, K. H. (2008). Kinetics of gold nanoparticles in the human placenta. *Reproductive Toxicology, 26*, 130–137.

Nel, A. E., Madler, L., Velegol, D., Xia, T., Hoek, E. M. V., Somasundaran, P., Klaessig, F., Castranova, V., & Thompson, M. (2009). Understanding biophysicochemical interactions at the nano-bio interface. *Nature Materials, 8*, 543–557.

Oberdörster, G., Oberdorster, E., & Oberdorster, J. (2005). Nanotoxicology: An emerging discipline evolving from studies of ultrafine particles. *Environmental Health Perspectives, 113*, 823–839.

Pan, Y., Neuss, S., Leifert, A., Fischler, M., Wen, F., Simon, U., Schmid, G., Brandau, W., & Jahnen-Dechent, W. (2007). Size-dependent cytotoxicity of gold nanoparticles. *Small, 3*, 1941–1949.

Patra, H. K., Banerjee, S., Chaudhuri, U., Lahiri, P., & Dasgupta, A. K. (2007). Cell selective response to gold nanoparticles. *Nanomedicine-Nanotechnology, 3*, 111–119.

Sadauskas, E., Wallin, H., Stoltenberg, M., Vogel, U., Doering, P., Larsen, A., & Danscher, G. (2007). Kupffer cells are central in the removal of nanoparticles from the organism. *Particle and Fibre Toxicology, 4*, 10.

Salmaso, S., Caliceti, P., Amendola, V., Meneghetti, M., Magnusson, J. P., Pasparakis, G., & Alexander, C. (2009). Cell up-take control of gold nanoparticles functionalized with a thermoresponsive polymer. *Journal of Materials Chemistry, 19*, 1608–1615.

Sawosz, E., Grodzik, M., Lisowski, P., Zwierzchowski, L., Niemiec, T., Zielinska, M., Szmidt, M., & Chwalibog, A. (2010). Influence of hydro-colloids of Ag, Au and Ag/Cu alloy nanoparticles on the inflammatory state at transcriptional level. *B Vet I Pulawy, 54*, 81–85.

Semmler-Behnke M., Fertsch S., Schmid G., Wenk A., Kreyling W. (2007). Uptake of 1.4 nm versus 18 nm gold nanoparticles by secondary target organs is size dependent in control and pregnant rats after intertracheal or intravenous application. *EuroNanoForum 2007*, 102–104.

Shenoy, D., Fu, W., Li, J., Crasto, C., Jones, G., DiMarzio, C., Sridhar, S., & Amiji, M. (2006). Surface functionalization of gold nanoparticles using hetero-bifunctional poly(ethylene glycol) spacer for intracellular tracking and delivery. *International Journal of Nanomedicine, 1*, 51–57.

Singh, N., Manshian, B., Jenkins, G. J. S., Griffiths, S. M., Williams, P. M., Maffeis, T. G. G., Wright, C. J., & Doak, S. H. (2009). NanoGenotoxicology: The DNA damaging potential of engineered nanomaterials. *Biomaterials, 30*, 3891–3914.

Sokolov, K., Aaron, J., Mack, V., Collier, T., Coghlan, L., Gillenwater, A., Follen, M., & Richards-Kortum, R. (2003a). Vital molecular imaging of carcinogenesis with gold bioconjugates. *Medical Physics, 30*, 1539–1539.

Sokolov, K., Follen, M., Aaron, J., Pavlova, I., Malpica, A., Lotan, R., & Richards-Kortum, R. (2003b). Real-time vital optical imaging of precancer using anti-epidermal growth factor receptor antibodies conjugated to gold nanoparticles. *Cancer Research, 63*, 1999–2004.

Takahashi, S., & Matsuoka, O. (1981). Cross placental-transfer of Au-198-colloid in near term rats. *Journal of Radiation Research, 22*, 242–249.

Taylor, U., Garrels, W., Petersen, S., Barcikowski, S., Klein, S., Kues, W., Lucas-Hahn, A., Niemann, H., & Rath, D. (2010a). Development of murine embryos after injection of uncoated gold and silver nanoparticles. *Reproduction, Fertility and Development, 22*, 240–241.

Taylor, U., Klein, S., Petersen, S., Kues, W., Barcikowski, S., & Rath, D. (2010b). Nonendosomal cellular uptake of ligand-free, positively charged gold nanoparticles. *Cytometry. Part A, 77*(5), 439–446.

Taylor, U., Petersen, S., Barchanski, A., Mittag, A., Barcikowski, S., & Rath, D. (2010c). Influence of gold nanoparticles on vitality parameters of bovine spermatozoa. *Reproduction in Domestic Animals, 45*, 60–60.

Thomas, M., & Klibanov, A. M. (2003). Conjugation to gold nanoparticles enhances polyethylenimine's transfer of plasmid DNA into mammalian cells. *Proceedings of the National Academy of Science of the United States of America, 100*, 9138–9143.

Wang, Z. X., & Ma, L. N. (2009). Gold nanoparticle probes. *Coordination Chemistry Reviews, 253*, 1607–1618.

Wiwanitkit, V., Sereemaspun, A., & Rojanathanes, R. (2009). Effect of gold nanoparticles on spermatozoa: The first world report. *Fertility and Sterility, 91*, e7–e8.

Zhang, X. D., Wu, H. Y., Wu, D., Wang, Y. Y., Chang, J. H., Zhai, Z. B., Meng, A. M., Liu, P. X., Zhang, L. A., & Fan, F. Y. (2010). Toxicologic effects of gold nanoparticles in vivo by different administration routes. *International Journal of Nanomedicine, 5*, 771–781.

Zielinska A. K., Sawosz E., Grodzik M., Chwalibog A., & Kamaszewski M. (2009). Influence of nanoparticles of gold on chicken embryos' development. *Annals of Warsaw University of Life Sciences - SGGW, Animal Science*, 249–253.

Chapter 13
Ultrasound Activated Nano-Encapsulated Targeted Drug Delivery and Tumour Cell Poration

Dana Gourevich, Bjoern Gerold, Fabian Arditti, Doudou Xu, Dun Liu, Alex Volovick, Lijun Wang, Yoav Medan, Jallal Gnaim, Paul Prentice, Sandy Cochran, and Andreas Melzer

Abstract

Introduction: Recently, ultrasonic drug release has been a focus of many research groups for stimuli responsive drug release. It has been demonstrated that a focused ultrasound (FUS) beam rapidly increases the temperature at the focused tissue area. One potential mechanism of drug targeting is to utilize the induced heat to release or increase penetration of chemotherapy to cancer cells. The efficiency of targeted drug delivery may increase by using FUS beam in conjugation with nano-encapsulated drug carriers.

The aim of this study is to investigate the effect of heat and ultrasound on the cellular uptake and therapeutic efficacy of an anticancer drug using Magnetic Resonance Imaging guided Focused Ultrasound (MRgFUS).

Materials and Methods: Human KB cells (CCL-17 cells) were seeded into 96-well plates and heat treated at 37–55°C for 2–10 min. Cell viability was determined using the colorimetric MTT assay. The cells were also subjected to MRgFUS and the degree of cell viability was determined. These

D. Gourevich • D. Xu • D. Liu
Institute of Medical Science and Technology, University of Dundee, Dundee, UK

CapsuTech Ltd, Nazareth, Israel
e-mail: d.gourevich@dundee.ac.uk

B. Gerold (✉) • A. Volovick
Institute of Medical Science and Technology, University of Dundee, Dundee, UK

InSightec Ltd, Haifa, Israel
e-mail: bgerold@dundee.a.uk

F. Arditti • J. Gnaim • A. Melzer
CapsuTech Ltd, Nazareth, Israel

L. Wang • P. Prentice • S. Cochran
Institute of Medical Science and Technology, University of Dundee, Dundee, UK
e-mail: bgerold@dundee.a.uk

Y. Medan
InSightec Ltd, Haifa, Israel

E. Zahavy et al. (eds.), *Nano-Biotechnology for Biomedical and Diagnostic Research*, Advances in Experimental Medicine and Biology 733, DOI 10.1007/978-94-007-2555-3_13,

experiments were conducted using an ExAblate 2000 system (InSightec, Haifa, Israel) and a GE 1.5 T MRI system, software release 15.

Results: We have observed a significant decrease in human KB cell viability due to heat (>41°C) in the presence of Doxorubicin (DOX), in comparison with DOX at normal culture temperature (37°C). The synergistic effect of heat with DOX may be explained by several mechanisms. One potential mechanism may be increased penetration of DOX to the cells during heating. In addition, we have shown that ultrasound induced cavitation causes cell necrosis.

Discussion and Future work: Further investigation is required to optimize the potential of MRgFUS to enhance cellular uptake of therapeutic agents. A novel delivery nano-vehicle developed by CapsuTech will be investigated with MRgFUS for its potential as a stimuli responsive delivery system.

Acknowledgments: This work is supported by an EU FP7 Industrial Academia Partnership Pathway IAPP.

Keywords Cancer • Cell culture • HIFU • Targeted drug delivery • Ultrasound

13.1 Introduction

The basic concept of Targeted Drug Delivery (TDD) is the ability to direct a certain drug to specific regions, tissues or cells in the human body (Patri et al. 2005). Utilization of TDD will be especially beneficial in cancer treatments, where the existing chemotherapy drugs cause severe side effects and could lead to formation of other cancer types in healthy cells (Freeman and Mayhew 1986).

The application of TDD is currently realized via two main approaches – chemical modifications and non-invasive intervention (Pua and Zhong 2009). Our current work aims to combine these methods to achieve a novel TDD technique, exploiting the benefits of both methods.

The approach based on chemical modifications generally consists of the formation of encapsulated drug delivery vehicles including iron-oxide based nanoparticles (Figuerola et al. 2010), liposomes (Maestrelli et al. 2005), peptide based polymers and others.

In this study we focus on cyclodextrin (CD) (Loftsson et al. 2007) based polymers developed by CapsuTech Ltd., as novel drug delivery vehicles to supply anti-cancer drugs to desired sites.

CDs are sugar- based cyclic molecules with the 3D shape of a truncated cone (Stella and He 2008). While the outer surface of the cone is hydrophilic, the inner cavity is relatively hydrophobic, a fact that allows the CD to encapsulate various hydrophobic drugs and to enhance their solubility (Challa et al. 2005). The encapsulation mechanism is represented in Fig. 13.1.

CDs have properties that may address some of the problems encountered with other vehicles. Since the drug-CD complex is in rapid equilibrium state, the release of the drug is readily achieved and spontaneous, in comparison to liposomes for example, which require more robust release methodologies (Schroeder et al. 2009). CDs have been proven to be non-toxic (Brewster and Loftsson 2007) and they are not degraded in the digestive system, making CDs potentially suitable for colon-specific drug delivery (Uekama et al. 1997).

By far the most important benefit of CDs, however, is that they solubilize, stabilize and even mask odours and tastes (Cabral Marques 2010) of hydrophobic drugs in aqueous environments, without altering

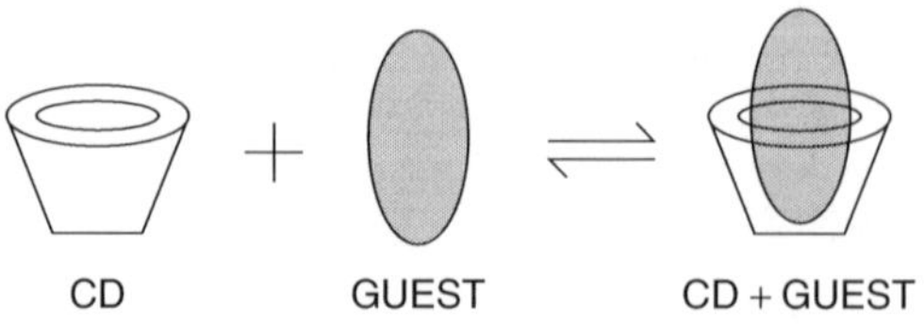

Fig. 13.1 Schematic representation of guest molecule encapsulation in CD to create a guest-CD complex (van de Manakker et al. 2009)

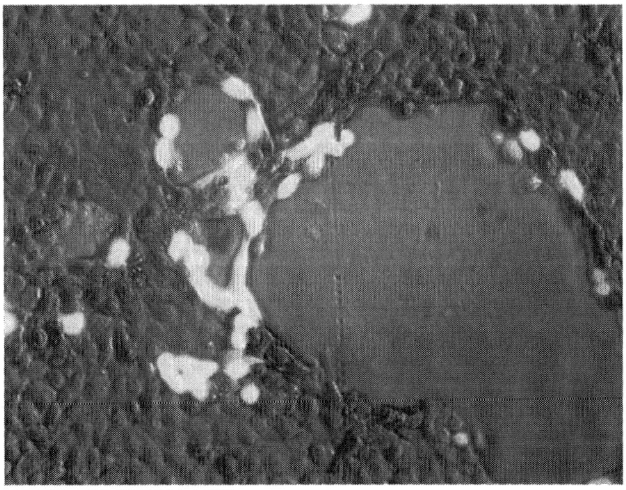

Fig. 13.2 A clearance zone in a cell monolayer due to cavitation (Prentice 2006)

their structure. Moreover, conjugation of CDs with other molecules may yield an even greater bio-effect. Combinations emerging from such conjugation include liposome-CD carriers, either by encapsulating the drug-CD complex inside liposomes, or by chemical modifications of the two carriers.

This conjugation of CDs can combine the CD's drug solubilisation effect with the liposomes ability to target drugs to certain locations. Other existing conjugations include CDs in micro/nanoparticles (Duchene et al. 1999), polymeric CD-based self-assembly of micelles and hollow spheres (Wang and Jiang 2006) and CD nanocarriers with Magnetic Resonance Imaging (MRI) contrast agents (Banerjee and Chen 2009; Battistini et al. 2008).

New, highly promising, non-invasive treatment modality is MRI guided Focused Ultrasound Surgery (MRgFUS) (Jolesz and McDannold 2008). MRgFUS combines high intensity focused ultrasound (HIFU) with high quality anatomical MRI images for treatment planning and real time thermometry monitoring, thus enabling controlled and completely non-invasive ablation of tumour tissue. MRgFUS has shown to be safe, reliable and effective therapeutic modality for non-invasive clinical interventions for various pathologies, such as ablation of uterine fibroids, bone metastasis palliative treatments, breast, liver and prostate cancer ablation.

Utilization of ultrasound (Tachibana and Tachibana 2001) as a trigger for targeted drug release (Aschkenasy and Kost 2005) and uptake increment (Hallow et al. 2006) is the subject of intense investigation by many research groups around the world. The drug release in specific locations is achieved via mechanical and/or thermal effects.

Sonoporation is a process that involves sonication of cells in order to modify the permeability of the cellular membrane. The permeability is affected by the creation of cavitation in the proximity to the cells. Cavitation refers to gas bubble activity, driven by application of the ultrasound (Pitt et al. 2006).

There are two distinguishable regimes of cavitation (Leightona 1994). The first type of cavitation is called stable (or non-inertial) cavitation, where the bubbles are oscillating, expanding at low ultrasound pressure and contracting at high ultrasound pressure. This typically occurs at lower intensities. These oscillations create a circulating fluid flow, often called micro-streaming, which generates shear forces on the membrane, thus increasing its permeability (Marmottant and Hilgenfeldt 2003). The second type of cavitation is unstable (also called inertial) cavitation, it occurs at higher intensity ultrasound exposures when the dynamic is dominated by the inertia of the liquid, and is associated with strong collapse phases and high localized energy densities. Near a boundary, the collapsing bubble can form a highly penetrating jet, which can punch a hole in the cell membrane, allowing the drug direct entrance to the cell, (Ohl et al. 2006) as demonstrated in Fig. 13.2.

In current clinical applications of FUS, cavitation is still considered a dangerous side effect, and generally avoided. A cavitation cloud formed in the focus of high intensity ultrasound field can grow rapidly with an unpredictable shape, especially at high ultrasonic amplitudes. Therefore, it is highly important to find a way to control the cavitation formation and its development in time. Studying of the cavitation phenomena will lead to an ability to control and actively use the cavitation in TDD via sonoporation phenomenon.

Thermosensitive liposomes (TSLs) (Dromi et al. 2007) are currently undergoing pre-clinical trials and are likely to be the first TDD agents to be used in patients. For this approach to TDD, HIFU is used as the source of hyperthermia, locally increasing the tissue temperature from 37°C up to 45°C degrees. During this process, the TSLs release their drug payload locally, within the volume of HIFU-induced hyperthermia (Staruch et al. 2011).

The research field of TDD advances rapidly, with constant novel developments; however, the need for specifically selective delivery vehicles, methods, devices and full process understanding still remains a very important factor in the TDD research. Especially the ability to detect and control cavitation creation will be a great step toward increased drug effect on damaged tissue, without hurting the healthy tissue.

In other words TDD will allow a monitored control of drug release *in situ*, simultaneously will reduce the given drug dose and the unwanted side effects, and eventually will lead to safer and more efficient treatments (Brouwers et al. 2009). The specific work reported here aims to achieve better understanding of the processes involved in ultrasound-activated TDD, such as heat, cavitation, encapsulation and others.

13.2 Materials and Methods

13.2.1 Cell Culture

The cancer cell lines used in this study were KB, which are considered to be HELA contaminated cells (ATCC CCL 17) (Hennache and Boulanger 1977; Akiyama et al. 1985) and A375m, human melanoma cells (ATCC CRL 1619) (Seftor et al. 1992; Middleton et al. 2000). Cells were cultured in Complete Medium (CM) consisting of RPMI 1640 medium supplemented with 10% fetal bovine serum (FBS), 1% penicillin streptomycin (5,000 I.U./ml, 5,000 lg/ml), 1% MEM Non-essential amino acids 100×, 0.5% HEPES, 1% Sodium Pyruvate 100 mM, all from Gibco Invitrogen, and 1% L-Glutamine 200 mM (Sigma-Aldrich). Dimethyl sulfoxide (DMSO) and Trypsin- EDTA 0.05% were obtained from Gibco Invitrogen. Cells were cultured as monolayers on 75 cm^2 cell culture flasks (TPP) in humidified air with 5% CO_2 at 37°C. Cell cultures were washed with PBS (Oxoid) and detached using trypsin EDTA.

For ultrasound exposure experiments, cells were seeded in an OptiCell growth chamber (50 cm^2 growth area) (NUNC), reaching confluence on the day of the experiment. The OptiCell (Glaser 2001) is a sterile growth environment treated for cell culture between two parallel gas-permeable, optically clear, polystyrene membranes surrounded by a standard plate-sized microtiter frame with two access ports. Doxorubicin was obtained from Mesochem Technology.

13.2.2 Cytotoxicity Measurements

The cytotoxicity of Doxorubicin to KB cells was measured by MTT (3-(4,5-Dimethylthiazol-2-yl)-2,5-diphenyltetrazolium bromide) (Sigma-Aldrich) – a colorimetric assay that determines the cell enzymatic activity. Cells were seeded in a 96-well tissue culture plate at 4,000 cells/well in 100 μL CM. Following the Doxorubicin insertion, cells were exposed to various heating protocols using an oven (Forced Air oven, DFO-36, MRC, Israel), the medium was changed and the cells incubated with fresh CM for 72 h.

At the end of the incubation period, 20 μL of MTT was added, and the cells were incubated for an additional 4 h, after which the liquid was removed and 100 μL of DMSO were added to dissolve the Formazan- an artificial chromogenic product created by the reduction of the tetrazolium salt (MTT). The Formazan absorbance was measured at 550 nm using a plate Elisa reader (Multiread 400, from Anthos-Labtec, Austria). Unheated cells were taken as control. Cell viability was calculated as following:

$$\text{Cell Viability} = (\text{Cell sample absorbance} / \text{control absorbance}) * 100$$

13.2.3 *Magnetic Resonance-Guided Focused Ultrasound Surgery (MRgFUS)*

The ExAblate 2000 system (InSightec, Israel) was used in a Magnetic Resonance Imaging (MRI) environment (1.5 Tesla GE, USA), for sonications. The focal location accuracy was verified before the experiment using tissue mimicking phantom and MRI thermometry. ExAblate 2000 possesses a built-in cavitation detector, based on spectrum measurement (Brennen 1995), which allows the user to identify cavitation occurrences.

13.2.4 *Cavitation Study*

To investigate the cavitation phenomenon in more details than a clinical MRgFUS system allows, we have developed an experimental set-up (Gerold et al. 2011), which makes possible observation of the first stages of the cavitation cloud formation in the high-speed regime. We have observed laser-nucleated acoustic cavitation with high speed cameras, at frame rates between 0.5 and 3 million frames per second.

13.3 Results

13.3.1 *Effect of Heat on Drug Uptake*

The effect of heating on the cytotoxicity and the uptake of Doxorubicin are represented in Figs. 13.3–13.5.

Figures 13.3–13.5 show a clear effect of temperature elevation on the cellular uptake of Doxorubicin and cells viability. As demonstrated both by bright-field and fluorescence images, there is enhanced uptake of Doxorubicin by cells attributable to a temperature increase of 8°C. In this manner, it is expected that HIFU-induced hyperthermia may enhance the local drug uptake.

13.3.2 *Cavitation*

Figure 13.6 illustrates cavitation cloud formation process. At time t=0 μs a laser pulse generates a nucleation site in a pre-established HIFU field. The size of the resulting cavitation cloud is very dependent on the intensity of the HIFU field. The HIFU field in Fig. 13.6a is of higher intensity than in Fig. 13.6b, yielding a larger cavitation cloud.

The ultrasound field presented in Fig. 13.6 was measured in terms of Mechanical Index (MI), which is defined as: $MI = \frac{PNP}{\sqrt{f}}$, where *PNP* is the peak negative pressure (MPa) and *f* is the frequency (MHz) of the ultrasound field.

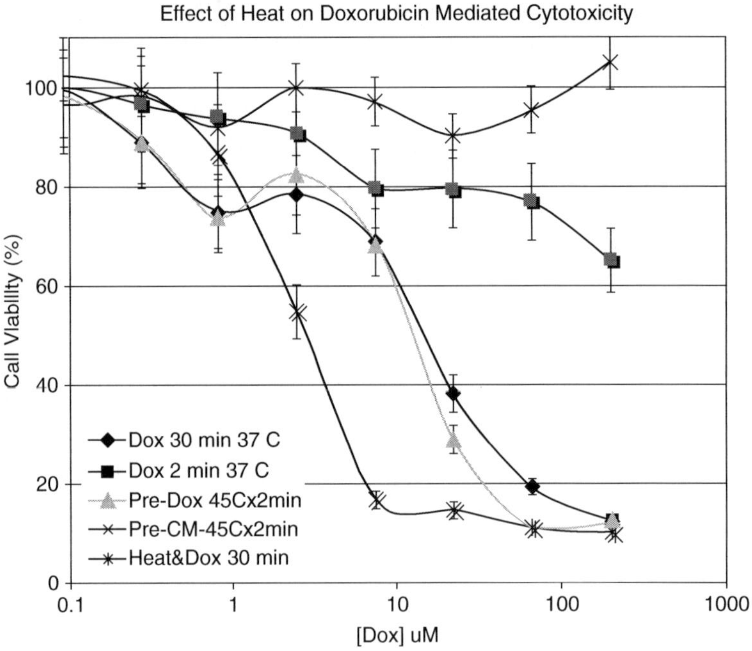

Fig. 13.3 The effect of temperature changes on the KB cell line viability in the presence of Doxorubicin, measured after 72 h by MTT assay

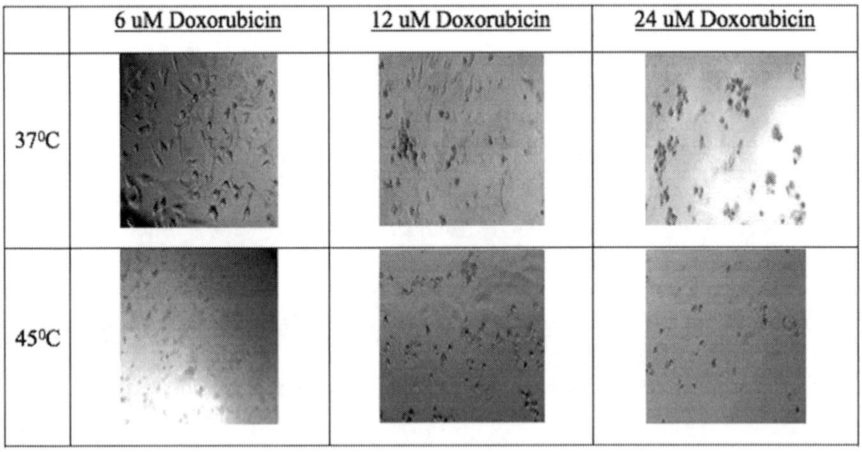

Fig. 13.4 KB cells response after 30 min exposure to various Doxorubicin concentrations, at two different temperatures. Magnification × 200

13.3.3 Cavitation with A375m Cells

Two different cell behaviors were observed according to the applied ultrasonic intensity (below and above the cavitation threshold). Figure 13.7 is typical example of sonications that did not produce cavitation (10 W of acoustic power with duration of 20 s) and did not have any observable effect on the cells. In contrast, sonications that produced cavitation (20 W of acoustic power with duration of 20 s) generated visible damage in the monolayers, such as displayed in Fig. 13.8. Cavitation detection relied on the ultrasound spectrum measurements by the ExAblate 2000 syste.

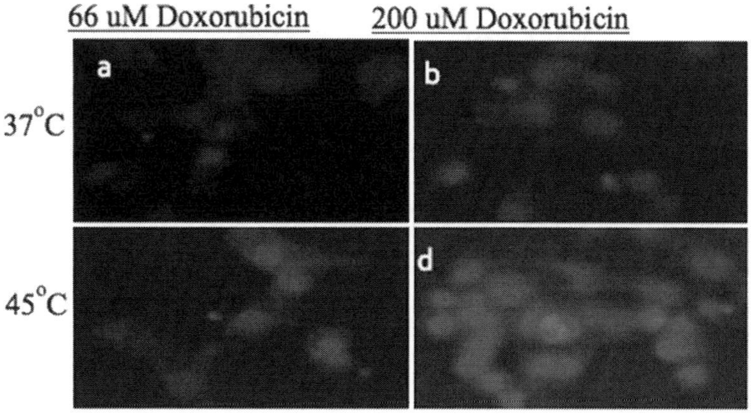

Fig. 13.5 Fluorescence images of Doxorubicin uptake by KB cell line after 2 min at different temperatures and concentrations. Magnification × 400

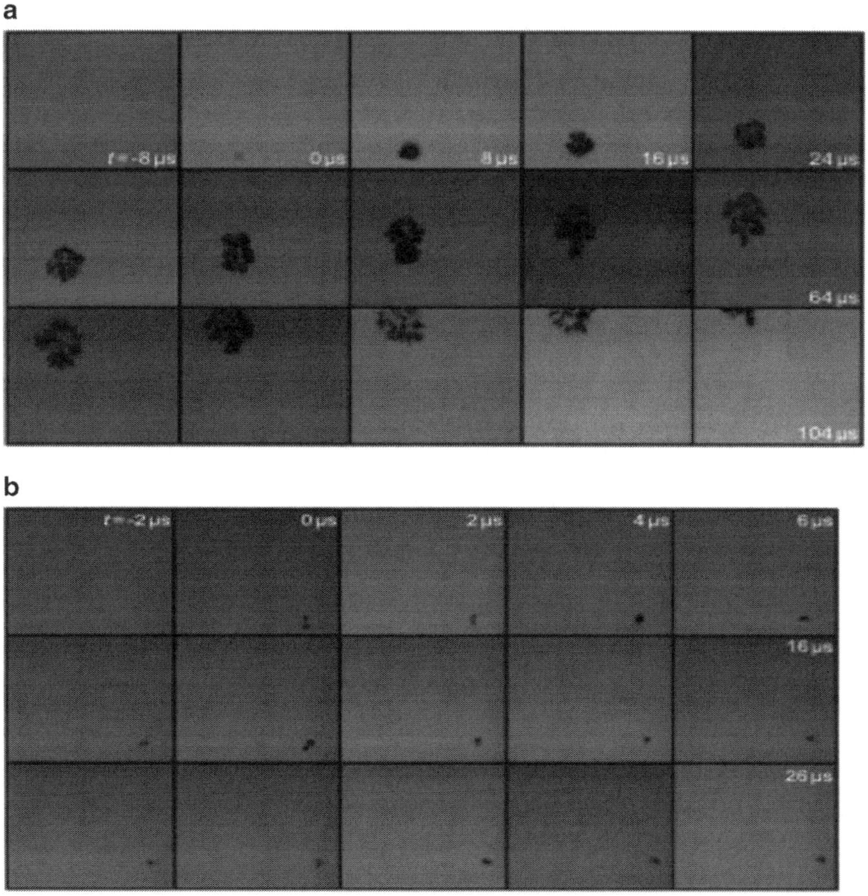

Fig. 13.6 Cavitation nucleation in a field of ultrasound mechanical index (**a**) MI = 3.4 and (**b**) MI = 1.7, recorded at 0.5 Mfps. The field-of-view for each image is 672 × 672 μm^2

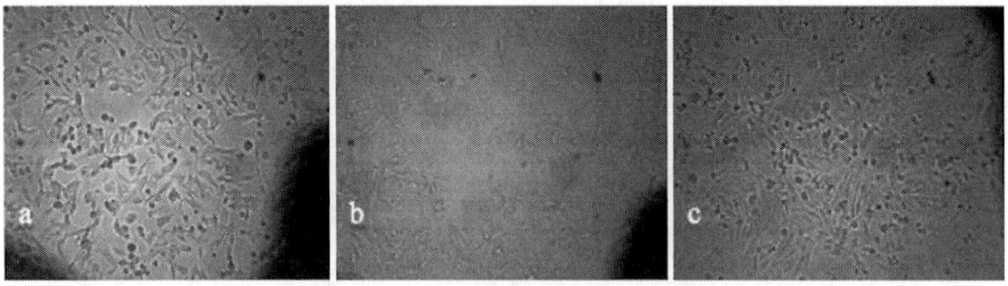

Fig. 13.7 The effect of sonication by ExAblate 2000 on A375m cells. Continuous sonication at 10 W for 20 s (**a**) before sonication, (**b**) immediately after sonication, (**c**) 24 h after sonication

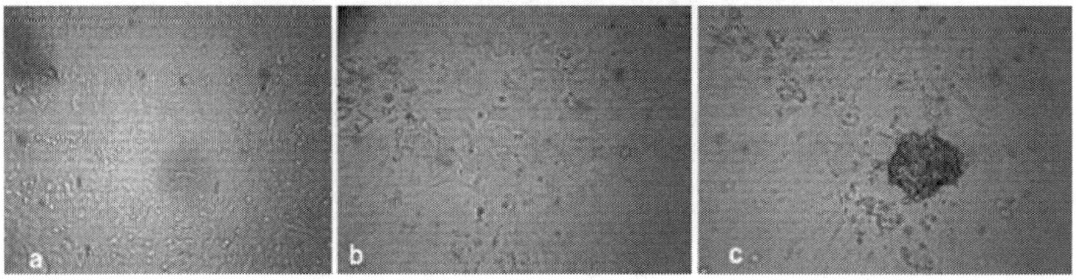

Fig. 13.8 The effect of ExAblate 2000 sonication on A375m cells. Continuous sonication at 20 W for 20 s (**a**) before sonication, (**b**) after sonication, (**c**) detached cell conglomerate near the sonication focus

13.4 Discussion

In this paper we have presented some preliminary results from our investigation of the use of high intensity focused ultrasound (HIFU) to mediate targeted drug delivery (TDD). The two potentially complementary mediation mechanisms that must be considered are sub-ablative heating and mechanical cavitation.

We have quantified the enhancement of Doxorubicin uptake into KB cells, at elevated temperatures similar to those that would be generated in a focused ultrasound field of intermediate intensity. Further work will employ ultrasound sonication to produce and maintain the thermal conditions we have identified.

In parallel, we have conducted FUS experiments on cell monolayers, with post-exposure analysis via optical microscopy. Consistent with the literature, we have observed enhanced molecular delivery to adherent cells, but the observations are dominated by large areas of cleared zones within the monolayer, at the location of the HIFU focal region.

We believe that cavitation, as detected during HIFU exposure, is acting to remove the cells from the surface on which they were previously attached. To investigate this further, we have begun cavitation studies, and are developing a technique, in which the location and moment of cavitation nucleation may be pre-defined. Crucially, this permits the use of ultra-high speed photography, which will greatly add to our understanding of the development and subsequent behavior of cavitation clouds during ultrasound exposure.

13.5 Future Work

We are currently working to bring the different areas of investigation together, to identify optimal conditions for TDD via HIFU for *in vivo* experiments, and ultimately for clinical trials. To achieve this, it is first necessary to gain a fundamental understanding of cellular bio-effects related to the physical mechanisms of heating and cavitation.

Acknowledgments This work is supported by an EU FP7 Industrial Academia Partnership Pathway IAPP.

References

Akiyama, S., et al. (1985). Isolation and genetic characterization of human KB cell lines resistant to multiple drugs. *Somatic Cell and Molecular Genetics, 11*(2), 117–126.

Aschkenasy, C., & Kost, J. (2005). On-demand release by ultrasound from osmotically swollen hydrophobic matrices. *Journal of Controlled Release, 110*, 58–66.

Banerjee Shashwat S., & Dong-Hwang Chen (2009). Cyclodextrin-conjugated nanocarrier for magnetically guided delivery of hydrophobic drugs. *J Nanopart Res, 11*, 2071–2078.

Battistini, E., et al. (2008). High-relaxivity magnetic resonance imaging (MRI) contrast agent based on supramolecular assembly between a gadolinium chelate, a modified dextran, and poly-β-cyclodextrin. *Chemistry – A European Journal, 14*, 4551–4561.

Brennen, C. E. (1995). *Cavitation and bubble dynamics*. New York: Oxford University Press.

Brewster, M. E., & Loftsson, T. (2007). Cyclodextrins as pharmaceutical solubilizers. *Advanced Drug Delivery Reviews, 59*, 645–666.

Brouwers, J., et al. (2009). Supersaturating drug delivery systems: The answer to solubility- limited oral bioavailability? *Journal of Pharmaceutical Sciences, 98*, 2549–2572.

Cabral Marques, H. M. (2010). A review on cyclodextrin encapsulation of essential oils and volatiles. *Flavour and Fragrance Journal, 25*, 313–326.

Challa, R., et al. (2005). Cyclodextrins in drug delivery: An updated review. *AAPS PharmSciTech, 6*(2) Article 43, p. E329–E357.

Dromi, S., et al. (2007). Pulsed-high intensity focused ultrasound and low temperature sensitive liposomes for enhanced targeted drug delivery and antitumor effect. *Clinical Cancer Research, 13*(9), 2722–2727.

Duchene, D., et al. (1999). Cyclodextrins in targeting application to nanoparticles. *Advanced Drug Delivery Reviews, 36*, 29–40.

Figuerola, A., et al. (2010). From iron oxide nanoparticles towards advanced iron-based inorganic materials designed for biomedical applications. *Pharmacological Research*. doi:10.1016/j.phrs.2009.12.012.

Freeman, A. I., & Mayhew, E. (1986). Targeted drug delivery. *Cancer, 58*, 573–583.

Gerold, B., et al. (2011). Laser-nucleated acoustic cavitation in focused ultrasound. *Review of Scientific Instruments, 82*(4), 044902.

Glaser, V. (2001). Current trends and innovations in cell culture. *Genetic Engineering, 21*(11).

Hallow, D. M., et al. (2006). Measurement and correlation of acoustic cavitation with cellular bioeffects. *Ultrasound in Medicine & Biology, 32*, 1111–1122.

Hennache, B., & Boulanger, P. (1977). Biochemical study of KB-cell receptor to adenovirus. *Biochemical Journal, 166*, 237–247.

Jolesz, F. A., & McDannold, N. (2008). Current status and future potential of MRI-guided focused ultrasound surgery. *Journal of Magnetic Resonance Imaging, 27*, 391–399.

Leighton, T.G. (1994). *The acoustic bubble*. Academic Press.

Loftsson, T., et al. (2007). Effects of cyclodextrins on drug delivery through biological membranes. *Journal of Pharmaceutical Sciences, 96*(10), 2532–2546.

Maestrelli, F., et al. (2005). Preparation and characterisation of liposomes encapsulating ketoprofen–cyclodextrin complexes for transdermal drug delivery. *International Journal of Pharmaceutics, 298*, 55–67.

Marmottant, P., & Hilgenfeldt, S. (2003). Controlled vesicle deformation and lysis by single oscillating bubbles. *Nature, 423*, 153.

Middleton, M. A., et al. (2000). Four-hourly scheduling of temozolomide improves tumour growth delay but not therapeutic index in A375M melanoma xenografts. *Cancer Chemotherapy and Pharmacology, 45*, 15–20.

Ohl, C. D., et al. (2006). Sonoporation from jetting cavitation bubbles. *Biophysical Journal, 91*(11), 4285–4295.

Patri, A. K., et al. (2005). Targeted drug delivery with dendrimers: Comparison of the release kinetics of covalently conjugated drug and non-covalent drug inclusion complex. *Advanced Drug Delivery Reviews, 57*, 2203–2214.

Pitt, W.G., et al. (2006). Ultrasonic drug delivery – A general review. *Expert Opinion on Drug Delivery*. Author manuscript; available in PMC 2006 February 6.

Prentice, P. (2006). *Membrane disruption by optically controlled cavitation*. PhD Thesis, University of Dundee.

Pua, E.C., & Zhong, P. (2009 Jan–Feb). Ultrasound-mediated drug delivery. *IEEE Engineering in Medicine Biology Magazine*, 64–75.

Schroeder, A., et al. (2009). Ultrasound, liposomes, and drug delivery: Principles for using ultrasound to control release of drugs from liposomes. *Chemistry and Physics of Lipids, 162*, 1–16.

Seftor, R.E.B., et al. (1992). Role of the alphaVbeta integrin in human melanoma cell invasion. *Proc. Natl. Acad. Sci. USA. Vol. 89*, 1557–1561.
Staruch, R., et al. (2011). Localised drug release using MRI-controlled focused ultrasound hyperthermia. *Int J Hyperthermia, 27*(2), 156–171.
Stella, Valentino J., & Quanren, He. (2008). Cyclodextrins. *Toxicologic Pathology, 36*, 30–42.
Tachibana, K., & Tachibana, S. (2001). The use of ultrasound for drug delivery. *Echocardiography, 18*, 323–328.
Uekama, K., et al. (1997). 6-O-[(4-Biphenylyl)acetyl]-α-,-β- and -γ-cyclodextrins and 6-Deoxy- 6-[[(4-biphenylyl) acetyl]amino]-α-,-β, and -γ-cyclodextrins: Potential prodrugs for colon- specific delivery. *Journal of Medical Chemistry, 40*, 2755–2761.
van de Manakker, F., et al. (2009). Cyclodextrin-based polymeric materials: Synthesis, properties, and pharmaceutical/ biomedical applications. *Biomacromolecules, 10*, 3157–3175.
Wang, J., & Jiang, M. (2006). Polymeric self-assembly into micelles and hollow spheres with multiscale cavities driven by inclusion complexation. *Journal of the American Chemical Society, 128*, 3703–3708.

Chapter 14
Ultrasound Mediated Localized Drug Delivery*

Stuart Ibsen, Michael Benchimol, Dmitri Simberg, and Sadik Esener

Abstract Chemotherapy is one of the frontline treatments for cancer patients, but the toxic side effects limit its effectiveness and potential. The goal of drug delivery is to reduce these side effects by encapsulating the drugs in a carrier which prevents release and can circulate throughout the body causing minimal damage to the healthy tissue. Slow release carriers have been developed which reduce the exposure to healthy tissue but this slow release also limits the maximum levels of drug in the tumor and nonspecific accumulation in healthy tissue remains a major hurdle. The next advance is to design these carriers to produce a rapid burst release of drug, but only in response to a localized trigger. The trigger of choice is low intensity focused ultrasound. A new particle is described here which incorporates an ultrasound sensitive microbubble of perfluorocarbon gas within a protective liposome carrier along with the payload. It is shown that this design can accomplish the desired burst release when exposed to ultrasound focused to small spatial locations within tissue phantoms. The ability to trigger release could provide a second level of spatial and temporal control beyond biochemical targeting or passive accumulation, making these promising particles for further development.

Keywords Burst release • Controlled release • Focused ultrasound • Microbubbles • Triggered drug delivery

---

* Stuart Ibsen and Michael Benchimol contributed equally to this work.

S. Ibsen (✉)
Department of Bioengineering, Moores Cancer Center, University of California San Diego,
3855 Health Sciences Dr. # 0815, La Jolla, CA 92093-0815, USA
e-mail: sibsen@ucsd.edu

M. Benchimol
Department of Electrical & Computer Engineering, Moores Cancer Center,
University of California San Diego, La Jolla, CA 92093, USA

D. Simberg
Moores Cancer Center, University of California San Diego, La Jolla, CA 92093, USA

S. Esener
Department of Nanoengineering, Moores Cancer Center,
University of California at San Diego, La Jolla, CA 92093, USA

E. Zahavy et al. (eds.), *Nano-Biotechnology for Biomedical and Diagnostic Research*,
Advances in Experimental Medicine and Biology 733, DOI 10.1007/978-94-007-2555-3_14,

14.1 Introduction

Chemotherapy is currently one of the major frontline treatments used for cancer patients. The drugs are usually injected systemically so that they circulate throughout the entire body. However, only a small fraction of the injected dose will actually reach the tumor and be therapeutic. The rest will circulate through the healthy tissue where it can cause detrimental side effects (Shapiro and Recht 2001). These side effects include hair loss and nausea and dangerous long term problems such as heart failure and the occurrence of new cancers. Chemotherapy drugs can also suppress the immune system which ultimately works against treating the patient. These side effects also limit the dose of drug that can be administered to a patient and often these limited doses fail to destroy the tumor. If chemotherapy is to advance as a treatment, these side effects must be reduced.

The drug delivery approach to reduce these side effects is to encapsulate the drug within a carrier. The carrier prevents release of the drug and can circulate through the body with very little effect on the healthy tissue. Certain drug delivery vehicles such as Abraxane for delivery of paclitaxel and liposomal Doxil for doxorubicin (Gabizon 2001; Miele et al. 2009) reduce exposure of non-targeted cells to the drug while accumulating a therapeutic dose within the tumor. Passive accumulation in the tumor tissue due to the enhanced permeation and retention of the vasculature (Gabizon 2001) coupled with slow drug release limits the bioavailability to non-tumor organs (Gabizon et al. 2003). However, this slow release also limits the maximum levels of drug in the tumor (Cheong et al. 2006), and non-specific accumulation in healthy tissue remains a major hurdle (Gabizon 2001).

To overcome these hurdles, the carriers can be designed to produce a rapid burst release of drug in response to a trigger. However, the carriers need to contain the drug without significant loss prior to trigger exposure. Designing carriers to have long term retention and controlled burst release is considered by many to be the next advancement in the drug delivery field. The ultimate success or failure of the design hinges on the choice of the trigger, which has several requirements. The trigger must change the carrier from a stable to an unstable state to initiate the release. The trigger must differentiate the tumor tissue from the healthy tissue and it must be tumor specific. These are challenging requirements to achieve when only using the biochemical or physical differences between the tumor and the healthy tissue because these differences often do not provide enough contrast (Vaupel et al. 1989). The use of tumor targeting ligands has the potential to improve the preferential accumulation of these delivery vehicles in tumor tissue (Karmali et al. 2009; Murphy et al. 2008). This delivery scheme requires endocytosis of the targeted vehicle with subsequent endosomal escape (Allen 2002; Ulrich 2002). However, saturation of the targetable receptors limits the number of vehicles that can bind to a cell. Also, tumor "receptors" are rarely unique to the tumor (Park et al. 2002) causing the targeted particles to accumulate in other healthy tissues, especially in the liver and spleen, causing local toxicity (Moghimi et al. 2001).

The use of enzymes, which are over expressed by the tumor, could be used to cleave bonds holding the carrier together, however these cleaving activities are usually performed by the liver as well (Rooseboom et al. 2004) and require the enzyme activity within the tumor to be significantly greater than that found in the healthy tissue for true tumor specificity. Tumors are also biochemically different from patient to patient (Fradet et al. 1987) which makes determining the properties of a specific patient's tumor a difficult and lengthy process.

The use of an artificial trigger can avoid many of the challenges encountered with biochemical triggers. One of the most promising artificial triggers is low intensity focused ultrasound. It can be used to specifically highlight the tumor tissue creating a stark differentiation between the tumor and the surrounding healthy tissue, a differentiation which is completely independent of biochemistry. One single carrier design which interacts with this trigger can be used in many different tumor types and in many different patients in a clinically relevant manner.

Focused ultrasound is a very attractive artificial trigger because it has been used for years in the biomedical imaging field. It is well characterized, and it is very safe as long as the exposure is kept below a threshold of 720 mW/cm^2 (Wojcik et al. 1995; Barnett et al. 2000). Focused ultrasound is capable of depositing useable amounts of energy into focal volumes of just a few cubic millimeters (Zanelli et al. 1993) deep within tissue. This allows it to be easily directed to desired locations in the body for pinpoint accuracy of carrier activation. These properties make focused ultrasound the trigger of choice.

However, safe intensity levels of ultrasound do not disrupt membranes. A particle which can operate as an ultrasound antenna is needed. These antenna particles must be very sensitive to ultrasound and can react to initiate an event. For years the most popular particles to achieve this effect have been microbubbles of gas (Stride and Saffari 2003) which have already been approved for human use as ultrasound contrast agents (de Jong et al. 2002; Von Bibra et al. 1999).

The density difference between the compressible gas and the surrounding water allows the microbubble to undergo size oscillations in response to the ultrasound pressure waves. This induces microstreaming of fluid around the microbubble and disrupts nearby membranes (Ferrara et al. 2007). If the properties of the microbubble match the driving frequency of ultrasound you can achieve resonance of the microbubble, resulting in a violent adiabatic implosion (cavitation) producing a shockwave and jets of water which can penetrate nearby cell membranes. These disturbances to the membrane can facilitate delivery of DNA or drugs into cells through a process known as sonoporation (Gao et al. 2007; Sun et al. 2006; Von Bibra et al. 1999; Zhao and Lu 2007).

Significant work has been conducted in an effort to use microbubbles as delivery vehicles in vivo (Gao et al. 2007; Zhao and Lu 2007). However, this has met with limited success (Willmann et al. 2008) due to extremely short circulation times of microbubbles in vivo (3–15 min half-life (Willmann et al. 2008)) and to limited payload capacity. Many delivery schemes use microbubbles by themselves as the drug carriers. When using just the microbubble itself you can load small amounts of hydrophilic payloads to the unprotected and limited surface area of the microbubble (Kheirolomoom et al. 2007; Klibanov 2006; Liu et al. 2006; Lentacker et al. 2006). You can also incorporate hydrophobic payloads in the limited volumes of thickened lipid, polymer, or oil surrounding the microbubble (Unger et al. 1998; Liu et al. 2006). However when these shells are fragmented, the hydrophobic drug will be contained in relatively large lipid particles reducing diffusion rates.

Other schemes use drug loaded liposomes as the carriers. Drug-loaded liposomes and microbubbles can be administered separately and targeted to the same tissue, but successful delivery of the drug depends on very close co-localization of both particles because the cavitation shockwave is only effective at disrupting membranes within a few tens of microns. It is unlikely that both particles would be present in sufficient proximity and concentration to deliver a therapeutic dose. To ensure colocalization, the drug loaded liposomes can be attached to the surface of microbubbles (Kheirolomoom 2007), however the points of attachment can concentrate shear stress during transport through the microvasculature and destabilize the entire particle.

To protect the microbubble and address the challenges described above, the microbubble and the payload can be encapsulated together within a protective outer liposome membrane shell as shown schematically in Fig. 14.1a. Our work demonstrates for the first time that these nested particles can be manufactured in a reproducible manner using a modified detergent dialysis process. The microbubbles of perfluorocarbon gas are first produced using traditional probe sonication methods and are subsequently encapsulated inside liposomes. We refer to these malleable nested structures as SHockwavE-Ruptured nanoPayload cArriers (SHERPAs). The actual SHERPA structure consisting of a nested 5 μm liposome containing a 1 μm microbubble is shown by fluorescence microscopy in Fig. 14.1b. The Brownian motion of the microbubble and payload of the SHERPA was contained entirely within the outer membrane as shown in a series of sequential pictures in Fig. 14.1c which demonstrates that the microbubble was encapsulated inside the structure. These structures were observed to be stable for several days.

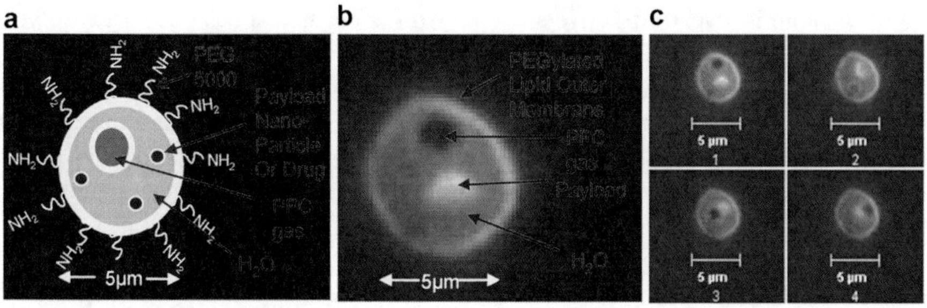

Fig. 14.1 *SHERPA nested structural design.* (**a**) Schematic representation of the nested liposome SHERPA design. (**b**) Fluorescent image of an actual SHERPA particle. The lipids have been labeled with a fluorescent dye to highlight the membranes. The payload is a small fluorescently labeled lipid membrane. (**c**) Confirmation that the microbubble is located within the liposome is shown in this series of sequential pictures showing the microbubble and fluorescent lipid payload moving around inside due to Brownian motion. The microbubble never left the confines of the outer membrane showing it was not on the surface

14.2 SHERPA Interaction with Ultrasound at an Intensity Level of 1.5 MPa

The interaction of these nested SHERPA structures with ultrasound was observed using high speed fluorescent videography on a custom built system. When these structures were exposed to focused ultrasound at 2.25 MHz the microbubble inside responded by undergoing a violent implosion and production of a cavitation shockwave as shown in Fig. 14.2. The SHERPA particle is designated by the red circle in frame 1. At the onset of ultrasound, shown in frame 2, the SHERPA particle is captured in a transitional state between being intact and undergoing cavitation. The cavitation itself occurred during the exposure of the frame resulting in a slightly blurred image between the two states. The resulting debris field is shown in frame 3 which consists of fragmented portions of the original liposome membrane and any contents from within the SHERPA. A jet of debris is pointed out by the arrow. These jets are created by asymmetric collapse of the microbubble itself and is a well documented mode of microbubble cavitation (Young 1999). Frame 4 shows the debris field beginning to diffuse away from the original site of cavitation. It is important to note the empty liposome in the lower left hand portion of frame 1. It does not contain a microbubble and serves as an important control because it was left completely unaltered by the ultrasound pulse. This shows that the ultrasound exposure was not high enough to cause damage to normal membranes. The observed SHERPA destruction occurred fully from the presence of the internal microbubble. This means that safe levels of ultrasound can be used to activate these SHERPA particles without causing damage to the membranes of surrounding cells and tissues. This localized effect of the cavitation on surrounding membranes illustrates the importance of the co-localization of the drug containing liposome and the microbubble.

14.3 SHERPA Interaction with Ultrasound at Intensity Levels Below 1 MPa

It was observed that even lower intensity levels of focused ultrasound could be used to activate SHERPA particles without inducing cavitation. Figure 14.3 shows the SHERPA particle in frame 1 before exposure to the low level ultrasound. The ultrasound was able to cause the microbubble to oscillate in size but not undergo cavitation. This oscillation created microstreaming of the fluid

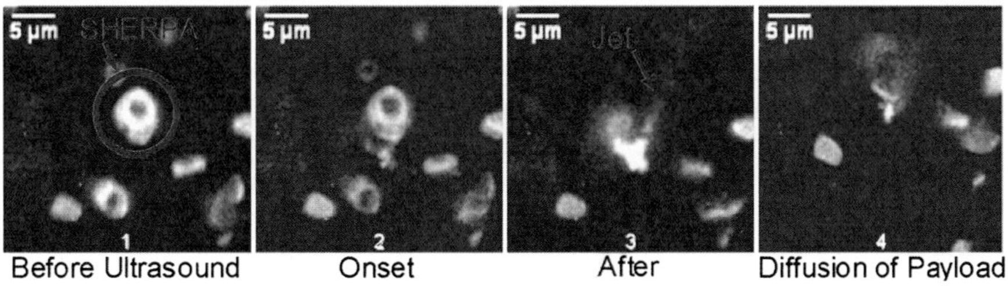

Fig. 14.2 *Interaction of SHERPA particles with focused ultrasound.* This sequence of images is taken from a high speed video and shows a cavitation mode of ultrasound interaction with a SHERPA. *Frame 1* shows the SHERPA before ultrasound exposure. *Frame 2* shows the very onset of ultrasound exposure. *Frame 3* shows the results just after the microbubble cavitation event creating a cloud of fluorescent debris. A jet of material has shot out from the main debris cloud. *Frame 4* shows the diffusion of the membrane fragments 1.2 s after the cavitation event

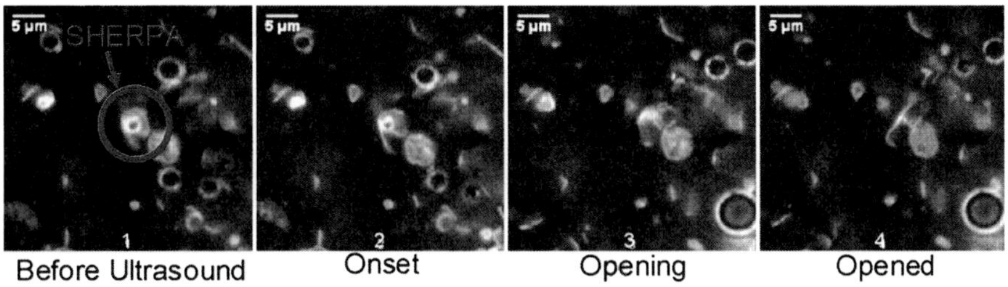

Fig. 14.3 *Interaction of SHERPA particles with lower intensity focused ultrasound.* This sequence of images is taken from a high speed video and shows a popping type mode of SHERPA interaction with lower intensity focused ultrasound. *Frame 1* shows the SHERPA with its fluorescent outer membrane before exposure to ultrasound. *Frame 2* shows the very onset of ultrasound exposure. *Frame 3* shows the SHERPA membrane popping open on the lower right hand side and beginning to open up. *Frame 4* shows the SHERPA fully opened up

surrounding the microbubble (Young 1999) and was in apart able to cause a rupture in the outer liposome membrane as seen in frame 2. Eventually the oscillating microbubble was destroyed by the pulse as shown in frame 3 but the outer liposome continued to open up and unfold. By frame 4 the outer membrane of the liposome was completely open. This shows that using ultrasound at intensities below the cavitation threshold can result in activation and opening of the SHERPA particle to release payload.

14.4 Localized SHERPA Activation

To obtain truly tumor specific activation of these SHERPA structures they must only rupture within a confined area around the focal zone of the ultrasound. The three dimensional localization of SHERPA activation was studied in an agar tissue phantom which simulated the acoustic properties of bulk tissue. A 1 mm diameter channel was molded through the center of the agar and was coated with avidin to simulate a blood vessel coated with targeting ligands. The SHERPA had biotin coating the inside which would attach to the avidin coated channels only if the SHERPA particle was ruptured. The internal biotin coating of the SHERPA was achieved by using DSPE-PEG2000-Biotin in the SHERPA manufacturing process and blocking any biotin that appeared on the surface of the SHERPA outer membrane by incubation with an excess of free avidin. The outer membrane was stained with DiI for visualization. The agar blocks were ensonified with focused ultrasound at different intensities.

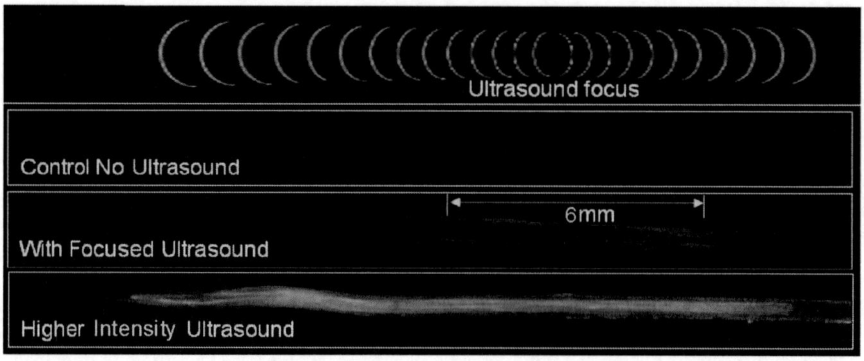

Fig. 14.4 *Localized effect of focused ultrasound on SHERPA particles.* The above three sets of images are from simulated blood vessel channels within agar tissue phantoms. The channels were coated with avidin and the SHERPA were functionalized with biotin where only free biotin was present on the inside of the SHERPA outer membrane. The control vessel was not exposed to ultrasound and showed very little nonspecific binding of the fluorescently labeled SHERPA to the surface of the channel. Ensonification with focused ultrasound ruptured SHERPA in the focal region exposing the biotin hidden on the inside and allowing it to bind to the surface of the channel. The use of higher intensity focused ultrasound activated more SHERPA particles and widened the range of SHERPA rupture causing larger deposition and brighter signal

The control agar block had the fluorescent SHERPA introduced into the channel and then washed out with water without any exposure to ultrasound. This showed a low level of nonspecific binding along the entire length of the channel as shown in Fig. 14.4.

The agar block that was exposed to low level ultrasound showed rupture of the SHERPA only in the focal region as shown by the increase in the presence of fluorescence along the channel walls present after washing. As the SHERPA particles opened up they exposed the unbound biotin from the inside and were able to stick to the walls.

The agar block exposed to higher intensity ultrasound showed a larger number of SHERPA ruptures which resulted in a higher fluorescence intensity signal as well as creating a larger region of activation.

14.5 Discussion

The ability to localize exposure of chemotherapy agents to just the tumor tissue has the potential to greatly reduce the systemic side effects experienced by cancer patients. The use of focused ultrasound is an attractive trigger to pinpoint drug release to just tumor tissue. However, there has historically been a lack of suitable drug delivery vehicles which have the ability to respond to focused ultrasound and release their payload.

The SHERPA particles described here, with their nested geometry, have several attractive features as ultrasound triggerable drug delivery vehicles. Their smooth continuous outer liposome with its PEG coating protects the internal microbubble and payload from degradation, reduces immune system recognition, and creates far greater loading capacity than microbubbles alone. Though the overall SHERPA structure is new, the outer liposome is amenable to standard functionalization and modifications well documented in the literature. These can increase their preferential accumulation in tumor sites to achieve maximum SHERPA concentration at the moment when the region is selectively ensonified with ultrasound. The surface area-to-volume ratio of these SHERPA is far less than that of nanoliposomes. This allows significantly more ligands on the cell surface to attach to a single SHERPA. This can increase targeting efficiency without the risk of receptor saturation and still deliver the same volume of drug. The lipid and cholesterol materials used in the construction of SHERPAs are bioresorbable and the perfluorocarbon gas can be cleared through exhalation. This prevents any dangerous permanent accumulation of materials within the body, which would ultimately limit the amount of SHERPA that could be injected over a lifetime. Unactivated SHERPA will break down gradually, diluting their payloads into the blood stream, making cellular delivery much less effective, and preventing accumulation of the drug (Ohl et al. 2006).

The main mode of administration for the SHERPA would be through systemic injection. The circulation of the SHERPA through the vasculature allows their payloads to penetrate the tumor wherever the vasculature reaches. This is a more efficient method to reach the advancing margins of the tumor as opposed to an intratumoral injection where the SHERPA would be limited in their mobility from the injection site. Although the center of the tumor would get a high dose of an injected sample, the tumor margins are of primary concern when looking to arrest tumor growth.

The SHERPA drug delivery vehicles themselves are not meant to extravasate from circulation into the tumor tissue. Their size prevents extravasation even though the flexibility of the outer membrane can mimic the flexibility of red blood cell membranes which helps to increase circulation time of the SHERPA by allowing easier passage through the microvasculature. However, once released from the SHERPA, the payloads consisting of nanoparticles and molecular drugs are capable of extravasation especially inside the tumor region due to the "leaky" vasculature. The main role of the SHERPA is to bring a highly concentrated payload into the tumor region through the vasculature. The endothelial cells of the tumor vasculature will be the first cells exposed to the burst release of payload from the SHERPA. The therapeutic payload can be chosen to kill these endothelial cells as well as attack the tumor cells themselves in an effort to reduce the vascularization of the tumor. Future work will explore the circulation time of these particles and the effect of focused ultrasound on payload delivery from the circulating particles to selected tissue regions.

The SHERPA also have the attractive feature of keeping the microbubble associated with the payload at all times. Not only does this guarantee that each particle will be activatable by the focused ultrasound it also creates the possibility of causing sonoporation in vivo. Sonoporation is the increase in cellular delivery through pore formation in the cell membrane near the microbubble cavitation event and has been shown in vitro as an effective method to improve delivery of particles into cells (Ohl et al. 2006). Transient holes formed in the cell membrane can be on the order of 100 nm in diameter and allow for payload uptake to occur over several minutes (Zarnitsyn et al. 2008). The benefit from the SHERPA particle is that the microbubble is always associated with the highly concentrated payload. The cavitation that would sonoporate the cell membrane would also cause simultaneous release of high concentrations of payload in the same region. This could allow payload to travel down its concentration gradient into the cells. This could create a new method for in vivo delivery, bypassing the need for endocytosis and endosomal escape as a means of intercellular delivery. This sonoporation effect may also occur during non-cavitation microbubble interactions from the microstreaming of fluid around the microbubble (Young 1999).

14.6 Conclusions

Low intensity focused ultrasound is a very attractive trigger to localize the release of payload from drug delivery vehicles. The localization of the drug release to just the tumor region should reduce the effect on the surrounding healthy tissue, helping to prevent systemic side effects. Here we introduce a new nested particle which has a microbubble encapsulated within a protective outer liposome. These particles have been demonstrated to be activatable with low levels of focused ultrasound. These ultrasound levels did not cause disruption to membranes of nearby empty liposomes. The activation of the particles has been shown to occur only near the focal zone of the ultrasound allowing for both spatial and temporal control over activation The resulting burst release of a highly concentrated payload from within the particle makes them promising for in vivo studies.

Acknowledgements The authors are grateful for the insightful discussions with Ahmet Erten while conducting these experiments. The study was supported by the NCI Grant No. 5U54CA119335-05, and by the UCSD Cancer Center Specialized Support Grant P30 CA23100.

References

Allen, T. (2002). Ligand-targeted therapeutics in anticancer therapy. *Nature Reviews Cancer, 2*, 750–763.

Barnett, S., Ter Haar, G., Siskin, M., Rot, H. D., Duck, F., & Maeda, K. (2000). International recommendations and guidelines for the safe use of diagnostic ultrasound in medicine. *Ultrasound in Medicine & Biology, 26*, 355–366.

Cheong, I., Huang, X., Bettegowda, C., Diaz, L. A., Jr., Kinzler, K. W., Zhou, S., & Vogelstein, B. (2006). A bacterial protein enhances the release and efficacy of liposomal cancer drugs. *Science, 314*, 1308–1311.

de Jong, N., Bouakaz, A., & Frinking, P. (2002). Basic acoustic properties of microbubbles. *Echocardiography, 19*, 229–240.

Ferrara, K., Pollard, R., & Borden, M. (2007). Ultrasound microbubble contrast agents: Fundamentals and application to gene and drug delivery. *Annual Review of Biomedical Engineering, 9*, 415–447.

Fradet, Y., Islam, N., Boucher, L., Parent-Vaugeois, C., & Tardif, M. (1987). Polymorphic expression of a human superficial bladder tumor antigen defined by mouse monoclonal antibodies. *Proceedings of the National Academy of Sciences of the United States of America, 84*, 7227–7231.

Gabizon, A. A. (2001). Pegylated liposomal doxorubicin: Metamorphosis of an old drug into a new form of chemotherapy. *Cancer Investigation, 19*, 424–436.

Gabizon, A., Shmeeda, H., & Barenholz, Y. (2003). Pharmacokinetics of pegylated liposomal doxorubicin: Review of animal and human studies. *Clinical Pharmacokinetics, 42*, 419–436.

Gao, Z., Kennedy, A. M., Christensen, D. A., & Rapoport, N. Y. (2007). Drug-loaded nano/microbubbles for combining ultrasonography and targeted chemotherapy. *Ultrasonics, 48*, 260–270.

Karmali, P. P., Kotamraju, V. R., Kastantin, M., Black, M., Missirlis, D., Tirrell, M., & Ruoslahti, E. (2009). Targeting of albumin-embedded paclitaxel nanoparticles to tumors. *Nanomedicine, 5*, 73–82.

Kheirolomoom, A., Dayton, P. A., Lum, A. F., Little, E., Paoli, E. E., Zheng, H., & Ferrara, K. W. (2007). Acoustically-active microbubbles conjugated to liposomes: Characterization of a proposed drug delivery vehicle. *Journal of Controlled Release, 118*, 275–284.

Klibanov, A. (2006). Microbubble contrast agents targeted ultrasound imaging and ultrasound assisted drug-delivery applications. *Investigative Radiology, 41*, 354–362.

Lentacker, I., de Geest, B., Vandenbroucke, R., Peeters, L., Demeester, J., de Smedt, S., & Sanders, N. (2006). Ultrasound-responsive polymer-coated microbubbles that bind and protect DNA. *Langmuir, 22*, 7273–7278.

Liu, Y., Miyoshi, H., & Nakamura, M. (2006). Encapsulated ultrasound microbubbles: Therapeutic application in drug/gene delivery. *Journal of Controlled Release, 114*, 89–99.

Miele, E., Spinelli, G. P., Miele, E., Tomao, F., & Tomao, S. (2009). Albumin-bound formulation of paclitaxel (Abraxane® ABI-007) in the treatment of breast cancer. *International Journal of Nanomedicine, 4*, 99–105.

Moghimi, S. M., Hunter, A. C., & Murray, J. C. (2001). Long-circulating and target-specific nanoparticles: Theory to practice. *Pharmacological Reviews, 53*, 283–318.

Murphy, E. A., Majeti, B. K., Barnes, L. A., Makale, M., Weis, S. M., Lutu-Fuga, K., Wrasidlo, W., & Cheresh, D. A. (2008). Nanoparticle-mediated drug delivery to tumor vasculature suppresses metastasis. *Proceedings of the National Academy of Sciences of the United States of America, 105*, 9343–9348.

Ohl, C.-D., Arora, M., Ikink, R., de Jong, N., Versluis, M., Delius, M., & Lohse, D. (2006). Sonoporation from jetting cavitation bubbles. *Biophysical Journal, 91*, 4285–4295.

Park, J., Hong, K., Kirpotin, D., Colbern, G., Shalaby, R., Baselga, J., Shao, Y., Nielsen, U., Marks, J., Moore, D., Papahadjopoulos, D., & Benz, C. (2002). Anti-HER2 immunoliposomes: Enhanced efficacy attributable to targeted delivery. *Clinical Cancer Research, 8*, 1172–1181.

Rooseboom, M., Commandeur, J. N. M., & Vermeulen, N. P. E. (2004). Enzyme-catalyzed activation of anticancer prodrugs. *Pharmacological Reviews, 56*, 53–102.

Shapiro, C., & Recht, A. (2001). Side effects of adjuvant treatment of breast cancer. *The New England Journal of Medicine, 344*, 1997–2008.

Stride, E., & Saffari, N. (2003). Microbubble ultrasound contrast agents: A review. *Proceedings of the Institution of Mechanical Engineers. Part H, Journal of Engineering in Medicine, 217*, 429–447.

Sun, Y., Zhao, S., Dayton, P. A., & Ferrara, K. W. (2006). Observation of contrast agent response to chirp insonation with a simultaneous optical-acoustical system. *IEEE Transactions on Ultrasonics, Ferroelectrics, and Frequency Control, 53*, 1130–1137.

Ulrich, A. S. (2002). Biophysical aspects of using liposomes as delivery vehicles. *Bioscience Reports, 22*, 129–150.

Unger, E., McCreery, T., Sweitzer, R., Caldwell, V., & Wu, Y. (1998). Acoustically active lipospheres containing paclitaxel: A new therapeutic ultrasound contrast agent. *Investigative Radiology, 33*, 886–892.

Vaupel, P., Kallinowski, F., & Okunieff, P. (1989). Blood flow, oxygen and nutrient supply, and metabolic microenvironment of human tumors: A review. *Cancer Research, 49*, 6449–6465.

von Bibra, H., Voigt, J. U., Froman, M., Bone, D., Wranne, B., & Juhlin-Dannfeldt, A. (1999). Interaction of microbubbles with ultrasound. *Echocardiography, 16*, 733–741.

Willmann, J. K., Cheng, Z., Davis, C., Lutz, A. M., Schipper, M. L., Nielsen, C. H., & Gambhir, S. S. (2008). Targeted microbubbles for imaging tumor angiogenesis: Assessment of whole-body biodistribution with dynamic micro-PET in mice. *Radiology, 249*, 212–219.

Wojcik, G., Mould, J., Lizzi, F., Abboud, N., Ostromogilsky, M., & Vaughan, D. (1995). Nonlinear modeling of therapeutic ultrasound. *1995 IEEE Ultrasonics Symposium Proceedings*, 1617–1622.

Young, F. (1999). *Cavitation*. London: Imperial College Press.

Zanelli, C. I., Demarta, S., Hennige, C. W., & Kadri, M. M. (1993). Beamforming for therapy with high intensity focused ultrasound (HIFU) using quantitative schlieren. *IEEE Ultrasonics Symposium*, 1233–1238.

Zarnitsyn, V., Rostad, C., & Prausnitz, M. (2008). Modeling transmembrane transport through cell membrane wounds created by acoustic cavitation. *Biophysical Journal, 95*, 4124–4138.

Zhao, Y. Z., & Lu, C. T. (2007). Recent advances in the applications of ultrasonic microbubbles as gene delivery systems. *Yao Xue Xue Bao, 42*, 127–131.

Chapter 15
Sonochemical Proteinaceous Microspheres for Wound Healing

Raquel Silva, Helena Ferreira, Andreia Vasconcelos, Andreia C. Gomes, and Artur Cavaco-Paulo

Abstract In this work, we report a novel approach using proteinaceous microspheres of bovine serum albumin (BSA), human serum albumin (HSA) and silk fibroin (SF) containing different organic solvents, namely *n*-dodecane, mineral oil and vegetable oil, to reduce the activity of human neutrophil elastase (HNE) found in high levels on chronic wounds. The ability of these devices to inhibit HNE was evaluated using porcine pancreatic elastase (PPE) solution as a model of wound exudates. The results obtained indicated that the level of PPE activity can be tuned by changing the organic solvent present on different protein microspheres, thus showing an innovative way of controlling the elastase-antielastase imbalance found in chronic wounds. Furthermore, these proteinaceous microspheres were shown to be important carriers of elastase inhibitors causing no cytotoxicity in human skin fibroblasts *in vitro*, making them suitable for biomedical applications, such as chronic wounds.

Keywords Elastase inhibition • Microspheres • Proteins • Sonochemistry • Wound healing

15.1 Introduction

A wound is the result of disruption of normal anatomic structure and function. Based on the nature of the repair process, wounds can be classified as acute or chronic wounds (Boateng et al. 2008; Lazarus et al. 1994). Acute wounds are usually tissue injuries that heal completely, with minimal scarring, within the expected time frame, usually 8–12 weeks. Chronic wounds, on the other hand, arise from tissue injuries that heal slowly, that is not healed beyond 12 weeks and often reoccur (Boateng et al. 2008; Lazarus et al. 1994). Acute wounds have low levels of protein-degrading enzymes, whereas exudates from non-healing chronic wounds contain elevated levels of proteases, like matrix metalloproteinases and HNE (Shapiro 2002; Trengove et al. 1996; Yager and Nwomeh 1999).

R. Silva (✉) • H. Ferreira • A. Vasconcelos • A. Cavaco-Paulo
Department of Textile Engineering, University of Minho, Campus de Azurém, 4800-058 Guimarães, Portugal

A.C. Gomes
Department of Biology, Centre of Molecular and Environmental Biology (CBMA), University of Minho, Campus de Gualtar, 4710-057 Braga, Portugal

E. Zahavy et al. (eds.), *Nano-Biotechnology for Biomedical and Diagnostic Research*, Advances in Experimental Medicine and Biology 733, DOI 10.1007/978-94-007-2555-3_15,

Thus, it has been postulated that lowering protease levels in the chronic wound to levels normally found in acute wounds may accelerate healing in the chronic wound (Lobmann et al. 2005; Yager and Nwomeh 1999; Edwards et al. 2004). Due to its involvement in such process, there is extensive literature exploring the different types of inhibitors or inhibitors formulations that may restore the normal levels of these enzymes in the above diseases (Edwards and Bernstein 1994). Nevertheless, several challenges remain that need to be taken into consideration in developing novel wound healing delivery formulations. Even so, there are few studies on the release of oils from protein microspheres to promote the inhibition of high levels of HNE found on chronic wounds exudate. The work presented here evaluated the capacity of three different organic solvents (*n*-dodecane, mineral oil and vegetable oil) to inhibit HNE in order to evaluate their ability to be used as a part of a wound management strategy. However, these organic solvents are insoluble in aqueous environments and would require a vehicle to promote the elastase inhibition in the wound environment. For the delivery of such compounds to the target area, proteinaceous devices, based on BSA, HSA and SF have been used. Serum albumin is one of the most extensively studied and applied proteins in this research field because of its availability, low cost, stability, unusual ligand binding properties and is recognized as the principal transport protein for fatty acids and others lipids that would otherwise be insoluble in the circulating plasma (Peters 1985; Spector et al. 1969; Yampolskaya et al. 2005). On the other hand, SF has a long history of use in clinical applications (Wang et al. 2007). This protein provides mechanical toughness, biocompatibility and biodegradability (Altman et al. 2003). In this work, microspheres are produced by means of ultrasonication of a two-phase starting mixture, consisting of protein solution (BSA, HSA or SF) and organic solvent.

15.2 Materials and Methods

15.2.1 Sonochemical Preparation of Proteinaceous Microspheres

Microspheres were synthesized by an adaptation of the Suslick method (Suslick and Grinstaff 1990) using the experimental set-up previously described (Silva et al. 2010). Briefly, the organic solvent (namely *n*-dodecane, mineral and vegetable oil) was added to the aqueous solution of proteins (5 $g\,L^{-1}$ for BSA and HSA; 10 $g\,L^{-1}$ for SF) in order to achieve a ratio of 60:40 (aqueous/organic phase). The bottom of the high-intensity ultrasonic horn (20 kHz) was positioned at the aqueous/organic interface employing an amplitude of 40% with a temperature of $10 \pm 1°C$ and with a total treatment time of 3 min. The separation of phases was accomplished in a few minutes because of the lower density of microspheres in relation to the water. Microspheres were then collected by centrifugation (2,000 g, 30 min) using the centricon tubes (molecular-weight cut-off of 100 kDa). Finally, the microspheres were sterilized with UV radiation for 2 h.

15.2.2 Characterization of Protein Microspheres Formation

The efficiency of microspheres formation was evaluated with the Lowry method, using BSA as standard. This method is based on the determination of protein in the supernatant. The efficiency of microspheres formation was calculated as follows:

$$\text{Microspheres formation}(\%) = \frac{[\text{Protein}]_i - [\text{Protein}]_f}{[\text{Protein}]_i} \times 100 \tag{15.1}$$

where $[\text{Protein}]_i$ and $[\text{Protein}]_f$ are the initial and the final protein concentration, respectively.

The zeta-potential and the size distribution of microspheres were determined at 25 ± 0.1°C using a Malvern zetasizer NS (Malvern Instruments) by electrophoretic laser Doppler anemometry and photon correlation spectroscopy (PCS), respectively. The proteinaceous microspheres were diluted 1:100 and the results were expressed as mean value ± standard deviation.

For STEM analysis, the diluted microspheres suspension was placed in Copper grids with carbon film 400 meshes, 3 mm diameter. The shape and morphology of microspheres were observed using a NOVA Nano SEM 200 FEI.

15.2.3 *Determination of PPE Activity Loss Over Time*

The activity of PPE was measured according to a method previously reported with some modifications (Tanaka et al. 1990). In brief, 30 μL of enzyme was mixed with 900 μL of reaction buffer 100 mM Tris-HCl, pH 8.0. The reaction was started with the addition of 70 μL of 4.4 mM of Suc-Ala-Ala-Ala-*p*-nitroanilide, a synthetic substrate for PPE. The reaction was carried out for 5 min, at 25°C, and the cleavage of the substrate was monitored spectrophotometrically at 410 nm. One unit is defined as the amount of enzyme that will hydrolyse 1.0 μmol of Suc-$(Ala)_3$-*p*NA per minute at 25°C, pH 8.0. To examine the inhibitory activity of the organic solvents, different concentrations of proteinaceous microspheres were added to a fixed amount of PPE solution. The incubation was carried out a 25°C and, at determined time points, aliquots were collected to monitor the decrease in elastase activity. Measurements were recorded in triplicate and the results were expressed as mean value ± standard deviation.

15.2.4 *Cytotoxicity Screening*

The proteinaceous microspheres were tested for cytotoxicity according to the ISO standards (10993-5, 2009). The BJ5ta cell line (normal human skin fibroblasts) was maintained according to ATCC recommendations (four parts DMEM containing 4 mM L-glutamine, 4.5 $g\,L^{-1}$ glucose, 1.5 $g\,L^{-1}$ sodium bicarbonate and one part of Medium 199, supplemented with 10% (v/v) of FBS, 1% (v/v) of Penicillin/Streptomycin solution and 10 μg mL^{-1} hygromycin B). The cells were maintained at 37°C in a humidified atmosphere of 5% CO_2. Culture medium was refreshed every 2–3 days. Cells were seeded at a density of 10×10^3 cells/100 μL/well on 96-well tissue culture polystyrene (TCPS) plates (TPP, Switzerland) in the day before experiments and then exposed to different microspheres concentrations added to fresh culture medium. At 24, 48 and 72 h of exposure, cell viability was determined using the alamarBlue® assay (Invitrogen, EUA). 10 μL of alamarBlue compound were added to each well containing 100 μL of culture medium. After 4 h of incubation at 37°C the absorbance at 570 nm, using 600 nm as a reference wavelength, was measured in a microplate reader (Spectramax 340PC). The quantity of resorufin formed is directly proportional to the number of viable cells. Data are expressed as means with standard errors of the means. Two-way ANOVA followed by post hoc Bonferroni test (GraphPad Prism 5.0 for Windows) was employed with statistically significant differences when $P < 0.05$.

15.3 Results and Discussion

15.3.1 Characterization of Proteinaceous Microspheres

The characterization of the microspheres obtained with these different proteins was focused in the study of microspheres formation yield, particle size, polydispersity index (PDI), zeta-potential and morphology. All the experiments were done with three different organic solvents and, while *n*-dodecane present 12 carbons linked by saturated bonds, mineral oil and vegetable oil present different degrees of saturated and unsaturated fats. Firstly, to determine the successful yield of protein that forms microspheres with different organic solvent, the Lowry procedure was used. Tables 15.1–15.3 show the results obtained for BSA, HSA and SF microspheres, respectively. The proteinaceous microspheres evidence a high yield on microspheres formation independently of the organic solvent used. However, these devices exhibit a slight difference of the mean size when vegetable oil and mineral oil were applied, relatively to the microspheres obtained with *n*-dodecane. It is known that certain properties of fatty acid residues in the molecule of triacylglycerol have significant effects on the fluidity of the oil. The linear "zig-zag" organization of single bonds enables the chains to be lined up close to each other and intermolecular interactions, such as van der Waals interactions, can take place promoting the size reduction, which explain the results obtained for *n*-dodecane. This system inhibits fluid flow, resulting in the relatively high viscosity of the oils. On the other hand, the presence of double bonds, can produce "kinks" in the geometry of the molecules. This prevents the chains becoming close to form intermolecular contacts, which results in an increased capability of the fluid to flow and, consequently, a highest size was achieved. This assay further suggests that the SF present the highest Z-average values when compared with BSA and HSA. According with our previous studies this is mainly due to the differences that exist in the macromolecular structures of proteins (Silva et al. 2011). The PDI values where similar, without significant changes, in the presence of different organic solvents and proteins. Moreover, there was no difference on the yield of microspheres formation, particle sizes and PDI, when the proteinaceous microspheres were prepared in PBS medium or in water. Another characteristic of polymeric microspheres that is of extreme interest is zeta-potential, which is a function of the surface charge of the particles (Attard et al. 2000). The proteinaceous microspheres, prepared with the three different organic solvents, present a negative charge on their surfaces (Tables 15.1–15.3). Negligible changes in zeta-potential values were observed between different organic solvents used. However, the results show that the zeta-potential value dropped significantly when protein microspheres are prepared with PBS solution (pH 7.4). Studies of current literature suggest that the zeta-potential is related to both surface charge and the local environment of the particle (composition of the surrounding solvent, the environmental pH value and ions in the suspension) (Zhang et al. 2008). Nevertheless, the zeta-potential is not an actual measurement of the individual molecular surface charge; rather, it is a measurement of the electric double layer produced by the surrounding ions in solution (i.e. counter ions). The effect of pH on the zeta-potential of the protein particles is explainable by considering that the surfaces of these materials contain pH-dependent ionisable functional groups, both acidic and basic, that can undergo dissociation and protonation (Berg et al. 2009). The results advise that the use of deionized water (H_2O, pH 5.5) augmented the negative charge on the microspheres surfaces.

Suspension stability is another important parameter for materials characterization in microemulsion research. Nevertheless, stability studies were performed over 4 months analysing the macroscopic aspect and measuring the size distribution and zeta-potential, in order to evaluate their physical stability. An increase in size is observed over time when aggregation of particles occurs. Size of proteinaceous devices did not otherwise change over 4 months. For a suspension system, zeta-potential is also an important index, which reflects the intensity of repulsive force among particles and the stability of dispersion. The reduced net charge on the surface and the accompanying reduction in repulsive forces between particles led to aggregation. However, after the time storage of proteinaceous

Table 15.1 Effect of organic phase (40%) on the yield value (%), Z-average (nm), PDI and zeta-potential (mV) on the formulation of BSA prepared in different dispersing medium (H_2O and PBS)

Organic phase	Yield (%)	Z-average (nm)	PDI	Zeta-potential (mV)	
				H_2O	PBS
n-dodecane	100±0	960±186	0.5±0.07	–43±2	–15±1
Mineral oil	100±0	1,203±141	0.6±0.02	–40±0	–16±1
Vegetable oil	100±0	1,382±551	0.5±0.04	–42±1	–18±1

Table 15.2 Effect of organic phase (40%) on the yield value (%), Z-average (nm), PDI and zeta-potential (mV) on the formulation of HSA prepared in different dispersing medium (H_2O and PBS)

Organic phase	Yield (%)	Z-average (nm)	PDI	Zeta-potential (mV)	
				H_2O	PBS
n-dodecane	99±1	1,117±111	0.4±0.01	–41±2	–16±1
Mineral oil	97±0	1,608±114	0.6±0.06	–39±2	–13±1
Vegetable oil	94±0	1,676±275	0.5±0.08	–42±4	–13±1

Table 15.3 Effect of organic phase (40%) on the yield value (%), Z-average (nm), PDI and zeta-potential (mV) on the formulation of SF prepared in different dispersing medium (H_2O and PBS)

Organic phase	Yield (%)	Z-average (nm)	PDI	Zeta-potential (mV)	
				H_2O	PBS
n-dodecane	98±1	1,505±193	0.6±0.07	–37±3	–12±0
Mineral oil	88±2	1,930±159	0.7±0.04	–39±2	–12±0
Vegetable oil	90±1	2,006±125	0.6±0.08	–39±2	–14±3

devices, no noticeable aggregates or obvious changes in zeta-potential were observed when microspheres were prepared in H_2O or PBS medium. The electrostatic repulsive forces present in these emulsions were sufficient to promote the stability over the time frame of storage. The use of electron microscopy can provide valuable information on particle size, shape and structure. STEM is an important technique to depict particle morphology and surface features. This technique demonstrated the spherical shape of proteinaceous particles (data not shown). The spherical shape would offer the highest potential for controlled release and protection of incorporated drugs, as they provide minimum contact with the aqueous environment, as well as the longest diffusion pathways. Comparing particles with any other shape, spherical particles also require the smallest amount of surface-active agent for stabilization, because of their small specific surface area (Bunjes 2005).

15.3.2 Inhibitory Activity

The major goal of this work is the development of proteinaceous devices with a specific biological functionality: the inhibition of elastase in chronic wounds. For this purpose, it was important to determine the inhibitory activity of the different organic solvents incorporated into protein devices. The highest concentration of proteinaceous microspheres used (5 $g\,L^{-1}$ for BSA and HSA and 10 $g\,L^{-1}$ for SF), were incubated with PPE. It was achieved more than 40% of activity loss for all the proteinaceous microspheres prepared with different solvents, after 5 min of incubation. Nevertheless, both mineral and vegetable oil demonstrated a higher activity loss than *n*-dodecane, especially when SF was used as carrier. This result can be due to the presence of monounsaturated and polyunsaturated fatty acids in different ratios on mineral and vegetable oil, when compared

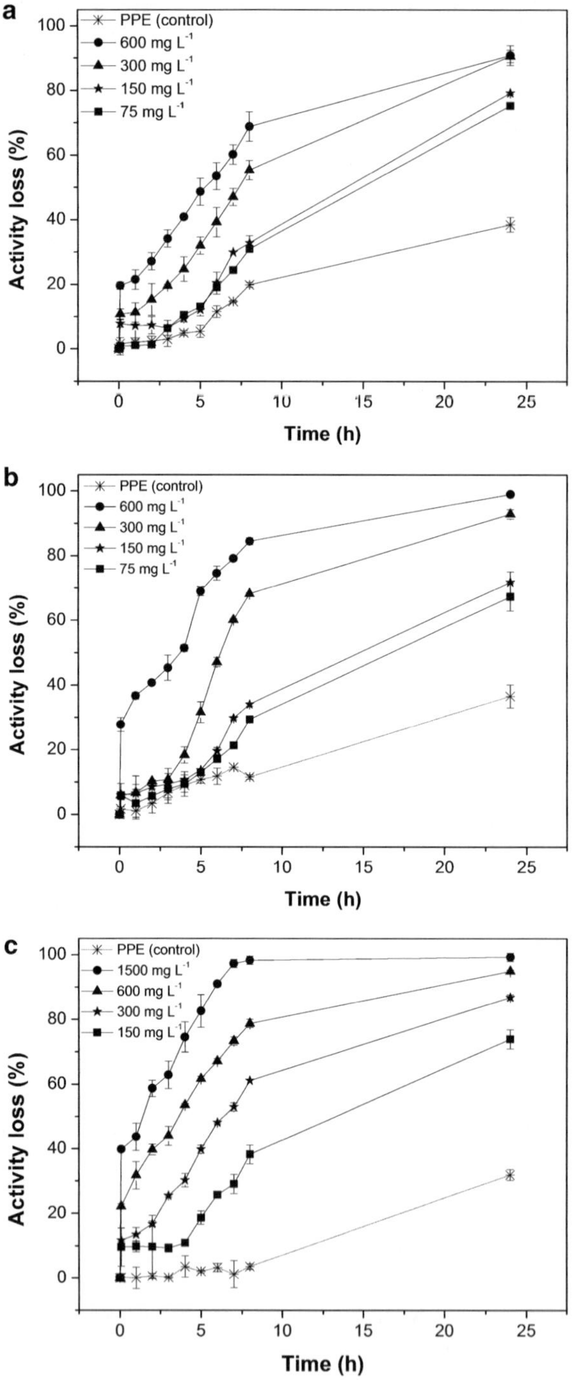

Fig. 15.1 Activity loss of PPE, obtained after 24 h of incubation, at 25°C, in the presence of different concentrations of BSA (**a**) 75, 150, 300 and 600 mg L^{-1}, HSA (**b**) 75, 150, 300 and 600 mg L^{-1} and SF (**c**) 150, 300, 600 and 1,500 mg L^{-1} microspheres incorporating vegetable oil

with *n*-dodecane, which is composed by saturated chains. Oleic acid, for example, has been shown to be effective in the wound healing process (Edwards et al. 2004). Considering all the mentioned aspects, the vegetable oil was used as standard conditions for all subsequent preparations. The controlled release of vegetable oil present in proteinaceous microspheres was also assessed over time. From the results obtained (Fig. 15.1) it is evident that for high protein microspheres concentrations,

PPE activity rapidly decreases, suggesting a promising system to modulate elastase activity. For lower protein microspheres concentrations, the decrease in activity of PPE was not so pronounced. This leads to the conclusion that the decrease in PPE activity is dependent on protein devices concentration, once that higher concentrations of proteinaceous microspheres lead to a higher content of the entrapped vegetable oil.

It is noteworthy that when SF is used as a carrier, a faster decrease of PPE activity is observed, revealing that the PPE solution is rapidly in contact with vegetable oil encapsulated into the SF microspheres. The data obtained can be related with the cristallinity degree of SF induced by ultrasound which was not sufficient to promote a more rigid structure, leading to an easier degradation or even the formation of cracks, allowing a rapidly diffusion out of vegetable oil. The results obtained for albumin microspheres can be related with their concentration in the devices. Previous reports indicated that albumin can increase enzyme activity (Edwards et al. 2004). Recently, the use of oleic acid/BSA formulations showed acceleration of the elastase activity, while requiring increased oleic acid concentrations to override the elastase activation properties of the albumin (Edwards et al. 2004). A possible explanation is that higher concentrations of BSA may stabilize the enzyme, by promoting aggregation or association of the albumin protein, which in turn may hinder the enzymatic turnover of available substrate. An alternative explanation is that the higher albumin concentration serves as an additional substrate for elastase hydrolysis and thereby decreases the level of enzyme available for chromophoric peptide substrate analysis (Edwards et al. 2004, 2007). These results further suggest differences between the two albumin proteins, bovine albumin and human albumin, in their capacity to inhibit PPE enzyme. The primary structures of HSA and BSA are homologous by 80% (Yampolskaya et al. 2005). The difference between these two proteins is mainly due to the replacement of some hydrophobic amino acids (a.a.) in their sequence and this replacement can promote different affinities to fatty acids (Yampolskaya et al. 2005; Peters 1985). Spector and co-workers (Spector et al. 1969) reported the affinity of BSA and HSA to several fatty acids, showing that the unsaturated fatty acids were bound more tightly to HSA than to BSA. This fact can explain the highest activity loss when HSA is used as a device.

15.3.3 Cytotoxicity Evaluation

The biocompatibility of proteinaceous devices incorporating the vegetable oil was assessed through the use of fibroblast *in vitro* cultures. A preliminary screening was made using murine embryonic fibroblasts cells (MEFs) as a measure of global cytotoxicity. In this study, a wide range of protein concentrations (from 75 to 5,000 mg L^{-1} for BSA, HSA and from 150 to 10,000 mg L^{-1} for SF) used for inhibition assays was evaluated and no toxicity was observed for the range concentrations of 75–300 mg L^{-1} BSA and HSA, and 150–600 mg L^{-1} for SF. The results obtained with MEFs were then confirmed with human skin fibroblasts (BJ5ta), testing the range protein microspheres concentrations that did not promote any damage to the MEFs. The effect of all tested conditions over BJ5ta was evaluated after 24, 48 and 72 h (Fig. 15.2).

Results indicate that neither protein solutions (BSA, HSA and SF), nor vegetable oil induced any toxicity in this time frame ($P>0.05$). It is possible to observe in Fig. 15.2 that the range concentrations of 75–300 mg L^{-1} of BSA and HSA microspheres containing the vegetable oil did not affect the viability of human skin fibroblasts cells for a period of 72 h, suggesting that microspheres can be safely applied to the skin at these concentrations. Similar data is displayed for the viability of BJ5ta cells, in the presence of HSA particles, which further provides the evidence of no cytotoxic effects after 72 h. However, the use of 600 mg L^{-1} of SF microspheres on human fibroblasts cells is statistically different when compared with control cells ($*P<0.05$), which suggests a mild cytotoxicity.

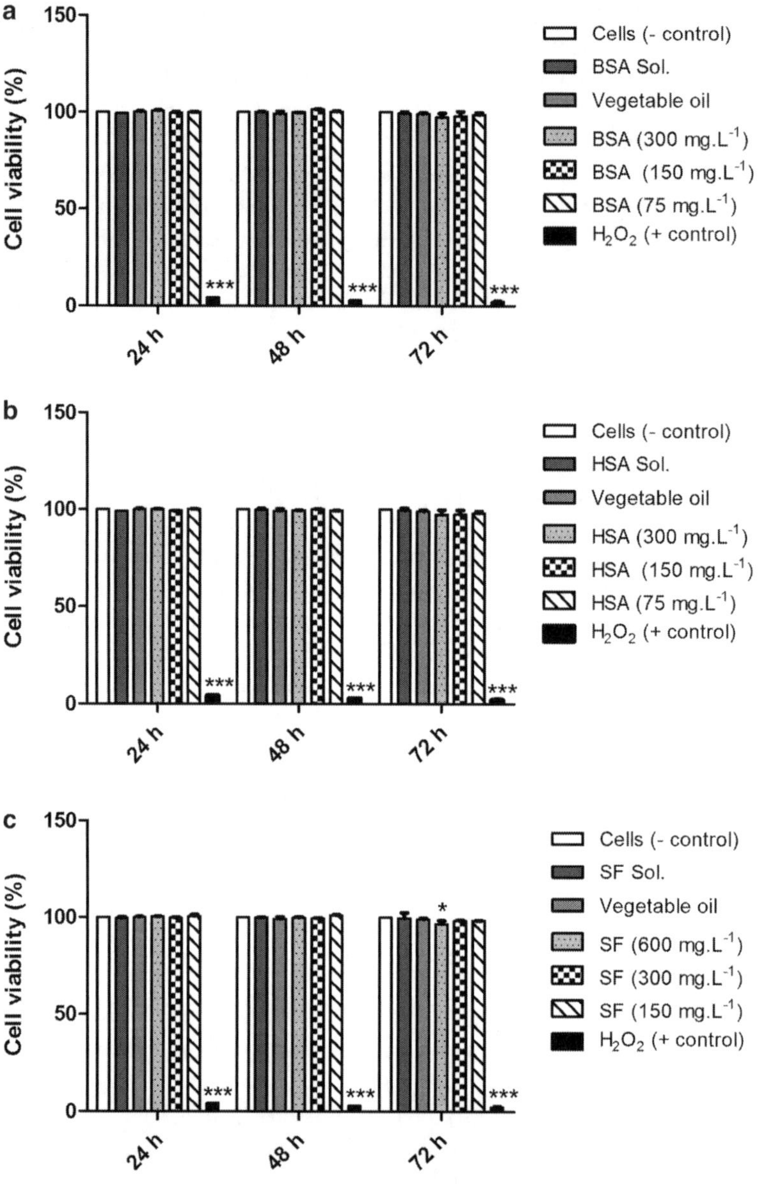

Fig. 15.2 BJ5ta cell viability at 24, 48 and 72 h of culture with different materials solutions and different concentration of BSA (**a**), HSA (**b**) and SF (**c**) microspheres incorporating vegetable oil. Values for tested samples are presented as a function of the control (cells cultured with culture medium, scored 100%). Statistically significant differences are observed. The results obtained were compared among each other and with the control: $*P<0.05$, $***P<0.001$

15.4 Conclusions

In this study, a promising microemulsion system was characterized and selected microemulsions were explored for their ability as potential elastase inhibitors.

This work outlines the feasibility of using different organic solvents to produce proteinaceous devices. After their production and subsequent characterization, the inhibitory activity was evaluated against PPE solution and it was found that vegetable oil presents higher inhibitory efficiency.

Furthermore, the effective nature of protein should be considered, once that SF demonstrated to be more capable of transferring vegetable oil in an aqueous solution under conditions of pH properties mimicking wound fluid. The cytotoxicity screening indicated that the developed protein particles are not cytotoxic in a wide range of concentrations (75–300 mg L^{-1} for BSA and HSA and 150–300 mg L^{-1} for SF) over 72 h of exposure. However, for those proteinaceous microspheres concentrations, the highest value attained for the inhibitory activity of PPE was achieved for HSA and BSA. Therefore, it is necessary to find a compromise between the ability to inhibit the elastase and their cytotoxicity, when formulations are developed. The results presented here strongly demonstrate the value of *in vitro* experiments with respect to the influence of proteinaceous devices to study single factors in the wound healing process. Nevertheless, it must be kept in mind that wound healing is rather complex, although the understanding of these properties may help to support the further refinement of protein devices for improved wound healing.

Acknowledgments We would like to acknowledge the financial support of European project Lidwine, contract no: NMP2-CT-2006-026741.

References

Altman, G. H., Diaz, F., Jakuba, C., Calabro, T., Horan, R. L., Chen, J., Lu, H., Richmond, J., & Kaplan, D. L. (2003). Silk-based biomaterials. *Biomaterials, 24*, 401–416.

Attard, P., Antelmi, D., & Larson, I. (2000). Comparison of the zeta potential with the diffuse layer potential from charge titration. *Langmuir, 16*, 1542–1552.

Berg, J. M., Romoser, A., Banerjee, N., Zebda, R., & Sayes, C. M. (2009). The relationship between pH and zeta potential of ~30 nm metal oxide nanoparticle suspensions relevant to in vitro toxicological evaluations. *Nanotoxicology, 3*, 276–283.

Boateng, J. S., Matthews, K. H., Stevens, H. N. E., & Eccleston, G. M. (2008). Wound healing dressings and drug delivery systems: A review. *Journal of Pharmaceutical Sciences, 97*, 2892–2923.

Bunjes, H. (2005). Characterization of solid lipid nanoparticles and microparticle. In C. Nastruzzi (Ed.), *Lipospheres in drug targets and delivery approaches, methods and applications*. Boca Raton: CRC Press LLC.

Edwards, P. D., & Bernstein, P. R. (1994). Synthetic inhibitors of elastase. *Medicinal Research Reviews, 14*, 127–194.

Edwards, J. V., Howley, P., & Cohen, I. K. (2004). *In vitro* inhibition of human neutrophil elastase by oleic acid albumin formulations from derivatized cotton wound dressings. *International Journal of Pharmaceutics, 284*, 1–12.

Edwards, J. V., Howley, P., Davis, R., Mashchak, A., & Goheen, S. C. (2007). Protease inhibition by oleic acid transfer from chronic wound dressings to albumin. *International Journal of Pharmaceutics, 340*, 42–51.

Lazarus, G. S., Cooper, D. M., Knighton, D. R., Margolis, D. J., Pecoraro, R. E., Rodeheaver, G., & Robson, M. C. (1994). Definitions and guidelines for assessment of wounds and evaluation of healing. *Archives of Dermatology, 130*, 489–493.

Lobmann, R., Schultz, G., & Lehnert, H. (2005). Proteases and the diabetic foot syndrome: Mechanisms and therapeutic implications. *Diabetes Care, 28*, 461–471.

Peters, T. J. (1985). Serum albumin. *Advances in Protein Chemistry, 37*, 161–245.

Shapiro, S. D. (2002). Proteinases in chronic obstructive pulmonary disease. *Biochemical Society Transactions, 30*, 98–102.

Silva, R., Ferreira, H., Little, C., & Cavaco-Paulo, A. (2010). Effect of ultrasound parameters for unilamellar liposome preparation. *Ultrasonics Sonochemistry, 17*, 628–632.

Silva, R., Ferreira, H., Araujo, R., Azoia, N., Rollet, A., Angel, U., Gomes, A., Freddi, G., Güebitz, G., Gedanken, A., & Cavaco-paulo, A. (2011) Insights on the mechanism of protein microspheres formation. Submitted

Spector, A. A., John, K., & Fletcher, J. E. (1969). Binding of long-chain fatty acids to bovine serum albumin. *Journal of Lipid Research, 10*, 56–67.

Suslick, K. S., & Grinstaff, M. W. (1990). Protein microencapsulation of nonaqueous liquids. *Journal of the American Chemical Society, 112*, 7807–7809.

Tanaka, H., Shimazu, T., Sugimoto, H., Yoshioka, T., & Sugimoto, T. (1990). A sensitive and specific assay for granulocyte elastase in inflammatory tissue fluid using L-pyroglutamyl-L-prolyl-L-valine-p-nitroanilide. *Clinica Chimica Acta, 187*, 173–180.

Trengove, N. J., Langton, S. R., & Stacey, M. C. (1996). Biochemical analysis of wound fluid from nonhealing and healing chronic Leg ulcers. *Wound Repair and Regeneration, 4*, 234–239.

Wang, X., Wenk, E., Matsumoto, A., Meinel, L., Li, C., & Kaplan, D. L. (2007). Silk microspheres for encapsulation and controlled release. *Journal of Controlled Release, 117*, 360–370.

Yager, D. R., & Nwomeh, B. C. (1999). The proteolytic environment of chronic wounds. *Wound Repair and Regeneration, 7*, 433–441.

Yampolskaya, G. P., Tarasevich, B. N., & Elenskii, A. A. (2005). Secondary structure of globular proteins in adsorption layers at the solution-air interface by the data of Fourier transform IR spectroscopy. *Colloid Journal, 67*, 385–391.

Zhang, Y., Yang, M., Portney, N., Cui, D., Budak, G., Ozbay, E., Ozkan, M., & Ozkan, C. (2008). Zeta-potential: A surface electrical characteristic to probe the interaction of nanoparticles with normal and cancer human breast epithelial cells. *Biomedical Microdevices, 10*, 321–328.

Chapter 16
Alendronate Liposomes for Antitumor Therapy: Activation of γδ T Cells and Inhibition of Tumor Growth

Dikla Gutman, Hila Epstein-Barash, Moshe Tsuriel, and Gershon Golomb

Abstract Circulating γδ T cells are cytotoxic lymphocytes that are unique to primates. Recent studies have shown that amino-bisphosphonates (nBP) activate γδ T cells to kill tumor cells in an indirect mechanism, which requires antigen presenting cells (APC). We hypothesized that selective targeting of nBP to monocytes would result in a more potent γδ T cells activation in circulation, and in tissue associated macrophages (TAM) following monocytes-laden drug extravasation and liposomes accumulation at the tumor site. In addition, inhibition of TAM by alendronate liposomes (ALN-L) is expected. ALN was targeted exclusively to monocytes, but not to lymphocytes, by encapsulating it in negatively-charged liposomes. The proportion of human γδ-T cells in the $CD3^+$ population following treatment with ALN-L or the free drug was increased, from 5.6 ± 0.4% to 50.9 ± 12.2% and 49.5 ± 12.9%, respectively. ALN solution and liposomes treatments resulted in an increased, and in a dose dependent manner, TNFα secretion from h-PBMC. Preliminary results showed that ALN-L inhibited tumor growth in a nude mouse breast tumor model. It is suggested that enhanced activation of γδ T cells could be obtained due to interaction with circulating monocytes as well as by TAM endocytosing liposomal nBP leading to a potentiated anti-tumor effect of nBP. It should be noted that this could be validated only in primates/humans since γδ T cells are unique in these species.

Keywords γδ T cells • Liposome • Monocyte • nBP • TAM

Abbreviations

DSPC	1,2-Distearoyl-sn-glycero-3-phosphocholine
ALN-L	Alendronate liposomes
nBP	Amino-bisphosphonate
APC	Antigen presenting cells
BSA	Bovine serum albumin
CLOD-L	Clodronate liposomes

D. Gutman (✉) • H. Epstein-Barash • M. Tsuriel • G. Golomb
Faculty of Medicine, Institute for Drug Research, School of Pharmacy,
The Hebrew University of Jerusalem, Box 12065, Jerusalem 91120, Israel
e-mail: diklag@ekmd.huji.ac.il

E. Zahavy et al. (eds.), *Nano-Biotechnology for Biomedical and Diagnostic Research*,
Advances in Experimental Medicine and Biology 733, DOI 10.1007/978-94-007-2555-3_16,

DCs	Dendritic cells
DSPG	Distearoyl-phosphatidylglycerol
MHC	Histocompatibility complex
h-PBMCs	Human peripheral blood mononuclear cells
imDC	Immature DC
IPP	Isopentenyl pyrophosphate
mAbs	Monoclonal antibodies
MPS	Mononuclear phagocytic system
TAM	Tissue associated macrophages

16.1 Introduction

Human circulating γδ T cells lymphocytes constitute of 1–5% of peripheral blood T cells (Hayday 2000) and exhibit major histocompatibility complex (MHC) unrestricted cytotoxicity against a large number of tumor types (Ferrarini et al. 2002). Most of the circulating γδ T cells belong to the Vγ9Vδ2 subset that are unique to primates (Hinz et al. 1997). Recent studies have shown that amino-bisphosphonates (nBP), antiresorptive drugs utilized clinically in bone-related disorders (Rodan 1998a), activate Vγ9Vδ2 T cells (Kunzmann et al. 2000). There has thus been interest in using nBPs in cancer immunotherapy, with promising results against B-cell malignancies (Wilhelm et al. 2003) and hormone refractory prostate cancer (Dieli et al. 2007). And in a very recent clinical trial, it was shown that a nBP (zolendronate) exerts a significant anticancer benefit when added to hormone therapy, reducing the risk of breast cancer returning by 36% (Gnant et al. 2009). Tumor cells are killed in an indirect mechanism, which requires antigen presenting cells (APC) (Caccamo et al. 2008; Kunzmann et al. 2000). nBP are internalized to some extent by mononuclear cells such as monocytes and dendritic cells (DCs) leading to intracellular accumulation of the isopentenyl pyrophosphate (IPP) metabolite (Roelofs et al. 2009). Consequently, this endogenous phosphoantigen is ultimately recognized by γδ T cells with subsequent cell activation, proliferation, and the release of TNFα, IL6 and IFNγ (Mariani et al. 2005). However, the family of nBP is both highly hydrophilic and charged, and is rapidly eliminated from the circulation by binding to bone and via urine excretion (Rodan 1998a, b; Fleisch 1998). Therefore, peripheral monocytes and tumors are exposed only briefly and to a relatively low concentration of nBP. Thus, in order to potentiate the effect of nBP selective targeting of the drug to APCs is necessary.

Encapsulating a BP in a particulate delivery system, such as liposomes, deviates these bone-seeking molecules to circulating monocytes and macrophages of the mononuclear phagocytic system (MPS) (van Rooijen and van Kesteren-Hendrikx 2003; Danenberg et al. 2002). The anti-inflammatory effect resulting from macrophage depletion by clodronate liposomes (CLOD-L; a non nBP) has been documented in experimental arthritis (Richards et al. 2001), delayed graft rejection (Slegers et al. 2000), CNS inflammation (Zito et al. 2001), and tumor angiogenesis (Zeisberger et al. 2006), and restenosis (Danenberg et al. 2002, 2003a; Epstein-Barash et al. 2010). Previous studies in our group demonstrated the high efficacy of alendronate liposomes (ALN-L; a nBP) in inflammatory-related disorders such as restenosis (Danenberg et al. 2002, 2003b; Epstein et al. 2007, 2008; Epstein-Barash et al. 2010) and endometriosis (Haber et al.. 2009, 2010).

Macrophages populate the microenvironment of most if not all solid tumors, representing >50% of the tumor mass in certain breast cancers (Lewis and Pollard 2006). Monocytes are recruited into solid tumor stroma where they differentiate into tumor-associated macrophages (TAM). Depletion of TAM by long-circulating CLOD-L, exploiting the enhanced permeability of the tumor microcirculation, has been reported (Banciu et al. 2008). A more potent activation of γδ T cells could be achieved by a liposomal delivery system due to the preferential uptake by circulating monocytes (Monkkonen et al. 1994; Monkkonen and Heath 1993; Epstein et al. 2008). The indirect therapeutic effect of BP in tumor is mediated by two different mechanisms, stimulatory effect on γδ T cells and the inhibition of TAM. We hypothesized that a potent therapeutic effect could be achieved by a liposomal delivery system of nBP, mediated by the two mechanisms. Selective targeting of nBP to monocytes by liposomes would result in a more potent γδ T cells activation in circulation, and at the tumor tissue by TAM following both, monocytes-laden drug extravasation and passive liposomes accumulation at the tumor site. In addition, inhibition of TAM by liposomal ALN is expected. Liposomal ALN is expected to be more potent than liposomal CLOD since this delivery system is more potent in depleting monocytes and macrophages (Danenberg et al. 2003a; Epstein-Barash et al. 2010). It should be noted that since ALN-L depleting effect on circulating monocytes is noted 36–48 h following treatment (Afergan et al. 2008; Epstein et al. 2007, 2008; Epstein-Barash et al. 2010), activation of γδ T cells can be achieved during this period.

Herein, we examined the effect of monocytes targeted ALN-L on γδ T cells activation and on tumor growth. To the best of our knowledge there are no reports in the literature exploiting circulating monocytes for the activation of γδ T cells by liposomal nBPs. Furthermore, the inhibitory effect of ALN-L on tumor growth has not yet been studied. It should be noted that the indirect mechanism of tumor inhibition mediated by γδ T cells activation couldn't be elucidated in the current study, because γδ T cells are unique to primates (Hinz et al. 1997).

16.2 Materials and Methods

16.2.1 Liposome Preparation

Liposomes were prepared by the modified thin-film hydration method (Danenberg et al. 2002; Epstein et al. 2008; Epstein-Barash et al. 2010) with 1,2-distearoyl-sn-glycero-3-phosphocholine (DSPC, Lipoid, Ludwigshafen, Germany), the negatively charged distearoyl-phosphatidylglycerol (DSPG, Lipoid), and cholesterol (Sigma-Aldrich, Israel). Liposomes were prepared at a molar ratio of 3:1:2. The lipids were dissolved in t-butanol and lyophilized overnight. The lyophilized cake was hydrated with an aqueous solution containing 200 mM ALN or CLOD at 55–60°C, and left to stand for 1 h. Blank liposomes were prepared by the same technique, with buffer instead of the drug. Fluorescent liposomes were prepared similarly with 0.025% w/w DSPE-rhodamine (Avanti Polar Lipids, Alabaster, AL) and 15.3% w/w dextran-FITC (Sigma-Aldrich, Israel), membrane and core markers, respectively (Afergan et al. 2008; Epstein-Barash et al. 2010). For small-sized liposomes, the ethanol injection method was used (Epstein et al. 2008; Epstein-Barash et al. 2010). The obtained liposomes were homogenized to a desired size by means of an extruder. To remove un-encapsulated drug, the liposomes were passed through a Sephadex G-50 column and were eluted with buffer.

16.2.2 Liposome Characterization

Liposome size and zeta potential were determined at room temperature after appropriate dilution with MES/HEPES buffer by photon correlation spectroscopy (ALV-GmBH, Langen, Germany) and NanoZ (Malvern Instruments, Malvern, UK), respectively. The phospholipid content was determined colorimetrically with the Bartlett assay (Bartlett 1959; Danenberg et al. 2002; Epstein et al. 2007), cholesterol content was measured by means of HPLC (Lang 1990), and ALN and CLOD content was determined by spectrophotometric assay of their complex with copper (II) ions at $\lambda = 240$ nm (Danenberg et al. 2003a; Epstein-Barash et al. 2010).

16.2.3 Ex Vivo Cell Culture

16.2.3.1 Human Peripheral Blood Mononuclear Cells (h-PBMCs)

h-PBMCs were isolated from human buffy coat using Ficoll-Hypaque (Amersham-Pharmacia, Uppsala, Sweden) density gradient centrifugation. h-PBMCs were cultured in RPMI media supplemented with 100 U/ml penicillin, 100 μg/ml glutamine, 10% fetal calf serum (Biological Industries, Beit Haemek, Israel), and 10 U/ml rhIL-2 (Chiron B.V., Amsterdam, The Netherlands).

16.2.3.2 Generation of Monocyte-Derived Immature DC (imDC)

ImDC were generated from monocytes, separated from adhered h-PBMC, by incubation in RPMI 1640 with 10% heat-inactivated human AB serum, supplemented with 1,000 U/ml IL-4 and 1,000 U/ml GM-CSF (R&D, Minneapolis, MN, USA) on days 1 and 4. ImDC were harvested on day five and phenotypically defined by FACS by the following monoclonal antibodies (mAbs): anti-CD14-PE (Dako, Glostrup, Denmark), anti-CD11C-FITC, anti-CD83-FITC and anti-HLA-DR-FITC (Beckman Coulter, Fullerton, CA, USA); and anti-86-PE (BD bioscience, San Jose, CA, USA). Isotype matched mAb were utilized as control (Beckman Coulter and BD bioscience).

16.2.4 Liposome Uptake by h-PBMC and imDC

Freshly isolated h-PBMCs and monocyte-derived imDC were incubated with fluorescent liposomes. Following 4 h of incubation, cells were harvested for assessment of liposomes cellular uptake by confocal microscopy (Zeiss LSM 410) and quantitative uptake by means of FACS (FACScan, Becton Dickinson, USA).

16.2.5 Quantification of TNFα Release

h-PBMCs were cultured at a density of $1 * 10^6$ cells/ml in 96-well u-shaped plates and treated with ALN solution or liposomes for 48 h in the presence of 10 U/ml rhIL-2. The medium was removed after centrifugation of the plate at 1,500 g, and levels of TNFα were quantified with ELISA kit (Peprotech EC Ltd, London, UK) according to the manufacturer's instructions.

16.2.6 Proliferation of γδ T Cells

h-PBMCs were cultured at a concentration of $1 * 10^6$ cell/ml in 24/well plates. Cells were treated for 9 days with BP solution or liposomes with or without 0.5 μg/ml LPS (LPS, E. coli, serotype 0127:B8, Sigma, Saint Louis, Missouri, USA) in the presence of 10 U/ml rIL-2. On day 4 and 7 of the culture period, one half of the medium was replaced with fresh medium containing rIL-2 and treatments. On day 9 non-adherent cells were harvested and aliquots of $1 * 10^6$ cells were washed with FACS buffer (1% w/v bovine serum albumin (BSA) and 0.02% w/v sodium azide in PBS pH 7.4) and stained with 10 μl anti-CD3-FITC and 5 μl anti-pan-γδ-TCR-PC5 Abs, or the respective isotype-matched controls

(Beckman Coulter, Fullerton, CA, USA) for 30′ at 4°C in a final volume of 100 μl FACS buffer. Adherent cells were detached using 0.25% trypsin/EDTA solution (Biological Industries, Beit Haemek, Israel) and aliquots of 1 * 10^6 cells were washed with FACS buffer and stained with anti-CD14 Ab (Dako, Glostrup, Denmark) for 30′ at 4°C in a final volume of 100 μl FACS buffer. Dead cells exclusion was done by adding 1 μg/ml topro3 (Invitrogen, California, USA) to a sample tube of both non-stained adherent and non-adherent cells. Cells were analyzed using BD FACS DIVA LSR II System (Becton Dickinson, USA) and FCS express (Denovo software) was used for quantitative analysis.

16.2.7 Murine Breast Cancer Model

Animal care and procedures were in accordance with the standards for care and use of laboratory animals of The Hebrew University of Jerusalem. Animals were fed with standard laboratory chow and tap water ad libitum. Female athymic nude mice (Harlan Laboratories, Jerusalem, Israel), 6 weeks old, were anaesthetized by isoflurane (Minrad International, USA), and were inoculated subcutaneously in the right flank with human MDA-231 (ATCC, Rockville, MD, USA) breast cancer cells (1 * 10^6 cells in 200 μl PBS). The injected cancer cells were allowed to grow for 14 days into tumors in the host animals before treatment. The animals were randomly assigned to a treatment of 20 mg/kg free ALN, 20 mg/kg ALN-L or empty liposomes (mean size of 85 ± 20 nm), by 3 lateral tail vain injections, on day 14, 16 and 18 (n = 4/group). The maximum and minimum diameters of the tumors were measured using a sliding caliper on day 14, 16 and 18. The volume of the tumors was calculated using the formula: 0.5(short axis)^2(long axis) (Geran et al. 1972).

16.2.8 Data Analysis

All data are expressed as the mean ± standard deviation. Comparisons among treatment groups were made by 2-way analysis of variance (ANOVA) followed by Tukey test, and unpaired two-tailed *t*-test when necessary. Differences were termed statistically significant at $p < 0.05$.

16.3 Results

16.3.1 Liposomes Characteristics

The physicochemical characteristics of ALN-L, used for γδ T cells activation and tumor growth inhibition, are summarized in Table 16.1. The liposomal formulations obtained were negatively charged and in two nano-sizes of 190 ± 24 and 85 ± 20 nm.

Table 16.1 Characterization of ALN-L formulations

Liposome composition (molar ratio)	ALN conc. (mg/ml)	Lipid conc. (mg/ml)	Mean diameter (nm)	Zeta potential (mV)
DSPC:Chol:DSPG (3:2:1)	5.6 ± 0.51	28.2 ± 2	190 ± 24	−30 ± 2.5
DSPC:Chol:DSPG (3:2:1)	5.2 ± 0.34	32.9 ± 1.7	85 ± 20	−31.5 ± 1.2

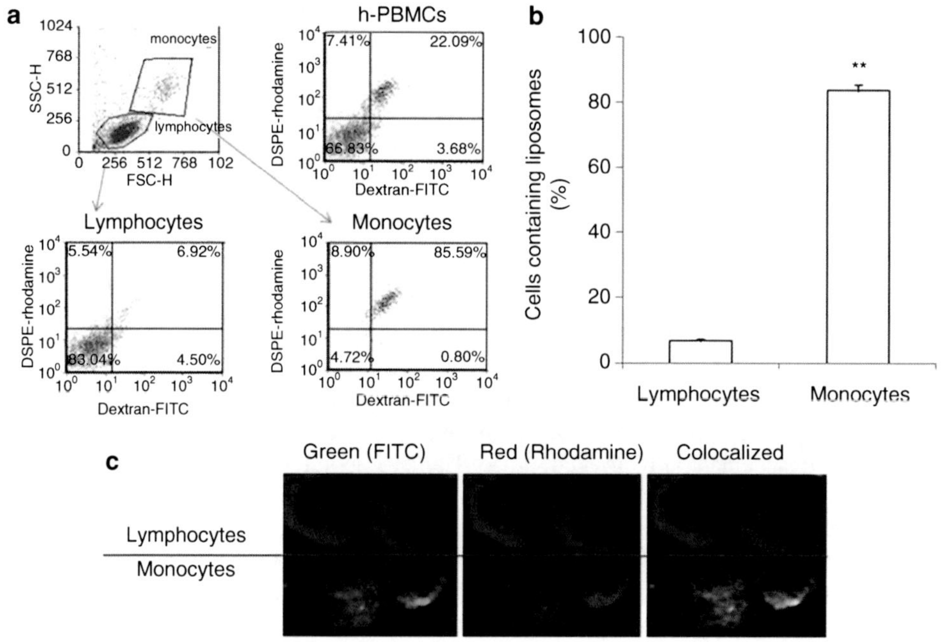

Fig. 16.1 Uptake of blank liposomes by h-PBMCs. Representative FACS images (**a**), calculated mean ± SD, N = 3 (**b**), and confocal images (**c**). h-PBMCs were incubated with fluorescent liposomes labeled with DSPE-rhodamine (membrane marker) and dextran-FITC (hydrophilic core marker). Following 4 h of incubation cells were harvested for assessment of liposomes internalization by confocal microscopy and FACS analysis (**$p < 0.01$)

16.3.2 Internalization of Liposomes by h-PBMCs Ex Vivo

In order to validate that the liposomes target specifically phagocytic cells, h-PBMCs, were incubated with fluorescent liposomes. The fluorescent liposomes were double-labeled with DSPE-rhodamine (red, membrane marker) and dextran-FITC (green, hydrophilic core marker). Following incubation with h-PBMCs, which comprise of monocytes and lymphocytes, the cells were analyzed by means of FACS. Treatment of h-PBMCs with blank liposomes resulted in internalization by 83 ± 1.8% of the monocyte cell population, whereas only 6 ± 0.6% of the lymphocytes internalized the liposomes (Fig. 16.1b). In agreement with the FACS results, confocal microscopy showed specific internalization of liposomes into monocytes (Fig. 16.1c, lower panel), and no liposomes were traced in association with lymphocytes (Fig. 16.1c, upper panel). Co-localization of the lipophilic membrane marker with the hydrophilic core marker (Fig. 16.1c right column, orange color) indicated the uptake of intact liposomes.

16.3.3 ALN Loaded Liposomes Bioactivity Ex Vivo

16.3.3.1 Proliferation of γδ T Cells

In cultures of h-PBMCs, the proportion of γδ T cells in the $CD3^+$ cell population was determined following various treatments in the presence of IL2 (Fig. 16.2). The control group, non-treated h-PBMCs, had 5.6 ± 0.4% γδ T cells in the $CD3^+$cell population. Treatment with ALN in solution or encapsulated in liposomes (1 μM) significantly increased the proportion of γδ T cells,

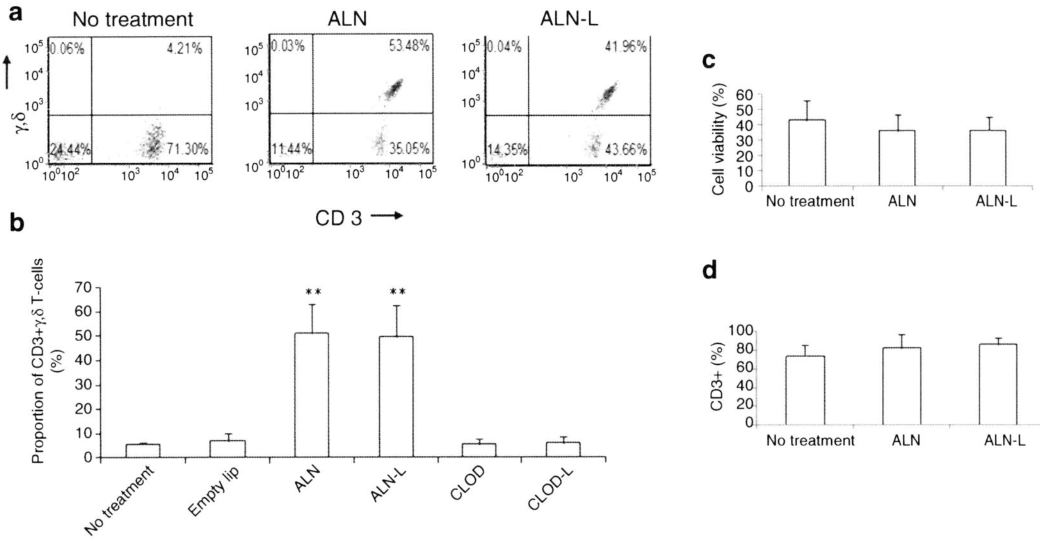

Fig. 16.2 The effect of bisphosphonates (clodronate, CLOD) and amino BP (alendronate, ALN) and liposomes (liposomal CLOD and ALN, CLOD-L and ALN-L, respectively) on γδ T cells proliferation. (**a**) FACS profiles of the T cells gated population. (**b**) Proportion of $CD3^+$ γδ-T cells (%). h-PBMCs were double stained with anti-CD3 and anti γδ TCR Abs before FACS analysis of the T-cell–gated population. (**c**) h-PBMCs viability by TO-PRO-3 staining and (**d**) $CD3^+$ counts. h-PBMCs were cultured for 9 days with 1 μM of various formulations in the presence of IL-2. Data shown is the mean ± SD of experiments with h-PBMCs from 4 to 7 independent donors (**$p<0.01$)

50.9 ± 12.2% and 49.5 ± 12.9% of the $CD3^+$cell population, respectively. Empty liposomes, CLOD (a non nBP) in solution and encapsulated in liposomes did not affect γδ T cells proliferation. The similar increase of the γδ T cell population following treatment with ALN as a free drug and in liposomes was associated with no changes in both cells' viability (Fig. 16.2c) and CD3 expression (Fig. 16.2d).

16.3.3.2 TNFα Secretion

After verifying the selective targeting of liposomes to monocytes in h-PBMCs cultures, the potential of the delivered ALN to exert its bioactivity on γδ T cell was studied by examining TNFα activation. Treatment with ALN in solution or in liposomes resulted in an increased, dose-dependent, TNFα secretion from h-PBMCs (Fig. 16.3).

16.3.3.3 Proliferation of γδ T Cells in LPS-stimulated Monocytes

Proliferation of γδ-T cells was evaluated in the presence of LPS-induced activation of monocytes. Elevated proportions of γδ-T cells were found in LPS treated h-PBMCs with or without ALN solution, in comparison with to no LPS, 56.4 ± 4.5%, 15.2 ± 3.2% and 44.6 ± 12.2%, 6.1 ± 3.4% of the $CD3^+$cell population, respectively. The stimulatory effect of ALN-L on γδ T cells proliferation was diminished when the h-PBMCs were co-treated with LPS, 52.5 ± 12% and 24.3 ± 4.8% of the $CD3^+$cell population, respectively (Fig. 16.4).

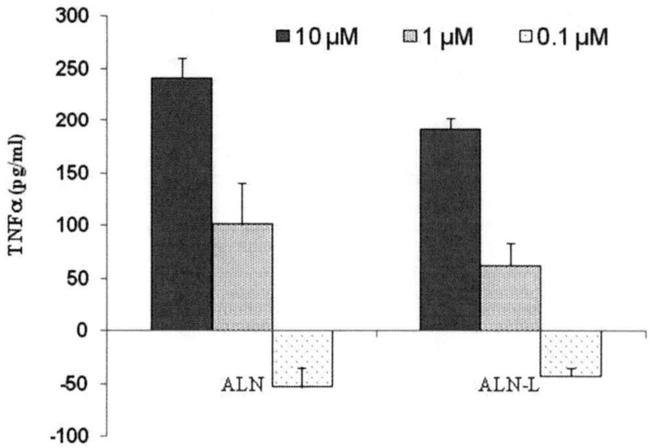

Fig. 16.3 The effect of ALN on TNFα secretion as a surrogate for γδ T activation. (**a**) h-PBMCs were incubated for 48 h with ALN in solution (ALN) or in liposomes (ALN-L) in the presence of IL-2. Data are values of h-PBMCs from two independent donors (mean ± SD)

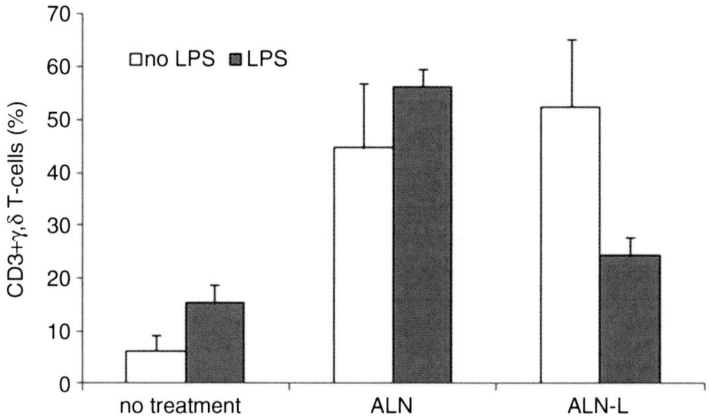

Fig. 16.4 The effect of alendronate (ALN) and liposomal ALN (ALN-L) on γδ T cells proliferation in the presence of LPS. The proportion of CD3+ γδ T cells was determined 9 days following h-PBMCs incubation with 1 μM ALN solution or liposomes in the presence of IL-2 with or without 0.5 μg/ml LPS. h-PBMCs were double stained with anti-CD3 and anti γδ TCR Abs before FACS analysis of the T-cell-gated population. Data shown is the mean ± SD of experiments with h-PBMCs from two independent donors

16.3.3.4 Interplay Between γδ T Cells and APCs

After assessing the bioactivity of ALN-L on γδ T cells, its effect on monocytes was further examined. Following treatment of h-PBMCs with ALN solution or liposomes, the proliferation of γδ T cells and the proportion of monocytes in the h-PBMCs culture were quantified (Fig. 16.5). In order to avoid monocytes depletion, a low concentration of ALN was used (<1 μM). Treatment with both ALN solution and liposomes resulted in elevated counts of γδ T cells in the $CD3^+$cell population, and in a dose dependent manner (Fig. 16.5a). The proliferation of γδ T cell was inversely correlated with $CD14^+$ expression in the adherent cells (Fig. 16.5b). It should be noted that lymphocytes and monocytes viability was similar in all treatments (data not shown).

Down regulation of the CD14 receptor could suggest differentiation of monocytes to dendritic cells (DC) following γδ T cell stimulation. In order to validate that liposomes also target DC, human imDC were incubated with fluorescently labelled liposomes. Human imDC were phenotypically defined as

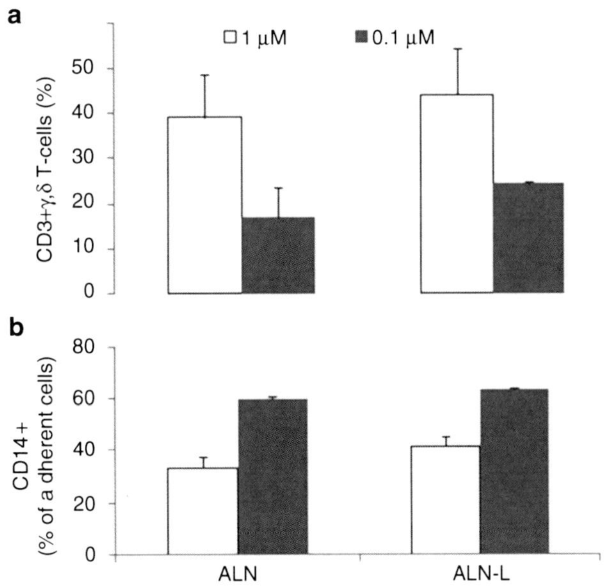

Fig. 16.5 The effect of alendronate (ALN) and liposomal ALN (ALN-L) on γδ T cells proliferation (**a**) and $CD14^+$ expression (**b**). h-PBMCs were cultured for 9 days with 0.1 and 1 µM ALN solution or liposomes in the presence of IL-2. Non-adherent cells were double stained with anti-CD3 and anti γδ TCR Abs and adherent cells were stained with CD14 Ab before FACS analysis. Data shown is the mean ± SD of experiments with h-PBMCs from two independent donors

$CD14^-$, $CD11c^+$, $CD83^-$, $CD86^+$ and HLA-DR^+ (Fig. 16.6 left). Following 4 h incubation with fluorescent liposomes, 48 ± 3% of the imDC, gated as $Cd11C^+$, were fluorescently stained (Fig. 16.6 right).

16.3.4 Effect of ALN-L on Tumor Growth In Vivo

The effect of ALN-L on tumor growth was studied *in vivo* (Fig. 16.7). Athymic nude mice were inoculated with human MDA-231 breast cancer cells, and blank liposomes, ALN and ALN-L were IV administered at day 14, 16 and 18. Animals treated with ALN solution exhibited necrosis at the injection area, and the animals were euthanized before the end of the experiment. This phenomenon was not observed in liposome treated animals, either empty or ALN loaded. Tumor growth inhibition was observed following treatment with ALN-L, but the results did not reach statistical significance due to the small number of animals.

16.4 Discussion

We demonstrate here that selective targeting of ALN to monocytes by conventional liposomes resulted in γδ T cells proliferation *ex vivo*. Furthermore, we show preliminary results of tumor growth inhibition *in vivo* by ALN-L in a mechanism not involving γδ T cells activation.

Activated γδ T cells display distinct natural killer functions and directly eliminate transformed cells, a feature that is successfully being exploited in immunotherapy trails in cancer patients (Dieli et al. 2007; Wilhelm et al. 2003). Intravenous stimulation by nBP of γδ T cells in patients for cancer immunotherapy is thought to involve accumulation of IPP in APCs (Eberl et al. 2009; Miyagawa et al. 2001; Roelofs et al. 2009). The full clinical impact of these drugs efficacy is impeded due to rapid urine elimination and bone accumulation (Rodan 1998a; Fleisch 1998). Moreover, due to their high charge and hydrophilicity, free BP do not easily cross cell membranes. We hypothesized that selective

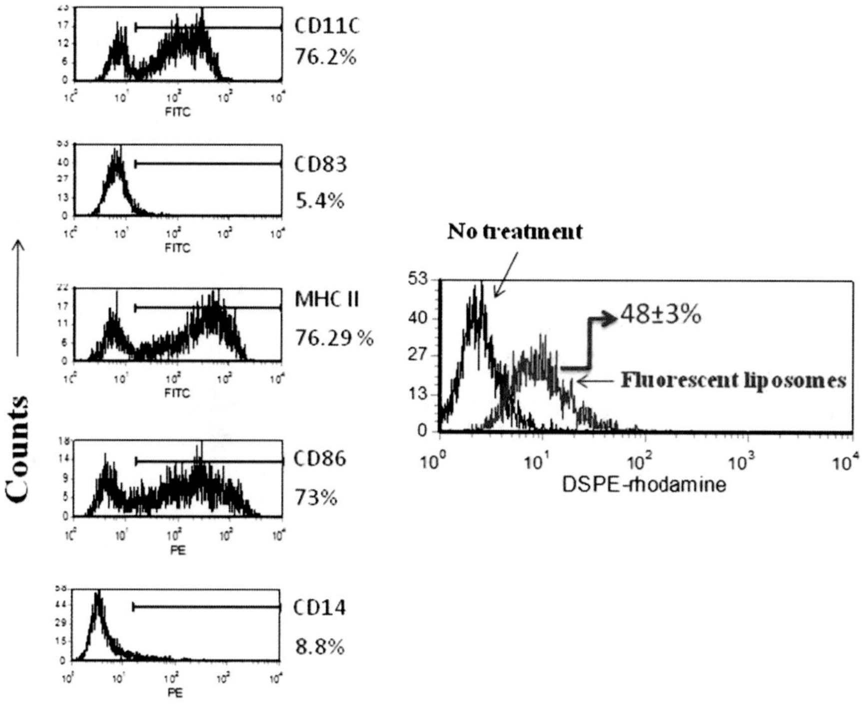

Fig. 16.6 Human derived imDC phenotype characterization and uptake of liposomes. FACS images of imDC phenotypes (*left*) and quantitative uptake of fluorescently labeled liposomes (*right*). imDC were generated from CD14$^+$ h-monocytes treated with GMCSF and IL4 for 5 days. imDC were incubated with fluorescent liposomes labeled with DSPE-rhodamine (membrane marker). Following 4 h of incubation cells were harvested for assessment of liposomes cellular uptake of CD11C gated cells (mean ± SD)

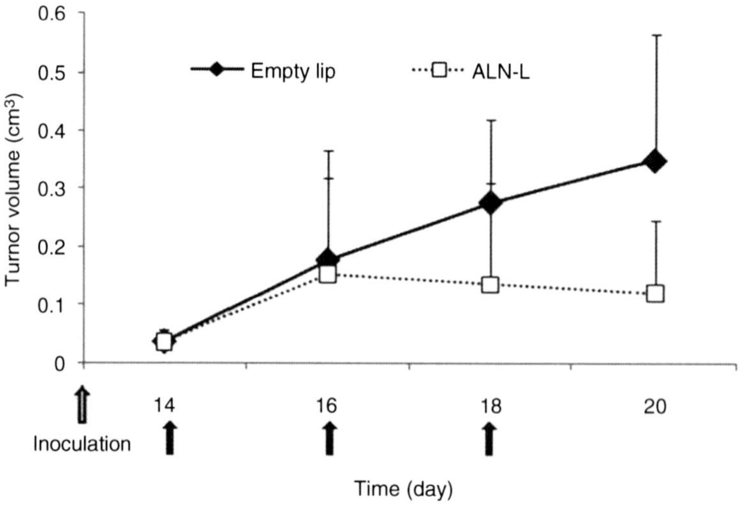

Fig. 16.7 The anti-tumor effect of liposomal alendronate (ALN-L) in athymic mouse model of human breast cancer. Shown is the tumor volume following treatment with ALN-L (20 mg/kg) and control (blank liposomes) by 3 IV injections (on day 14, 16, and 18) 14 days post tumor cells inoculation. Results are presented as the mean ± SD (n = 4)

targeting of nBP to monocytes would result in a more potent γδ T cells activation. We utilized liposomes to deliver encapsulated ALN to APCs for optimizing ALN efficacy due to increased uptake. Since ALN-L depleting effect on circulating monocytes is noted 36–48 h following treatment (Afergan et al. 2008; Epstein et al. 2007, 2008; Epstein-Barash et al. 2010), activation of γδ T cells can be achieved during this period. In this study we have used 'conventional' liposomes endowed with preferable physicochemical properties for monocyte-targeting; negatively charged membrane, that enhances the internalization of liposomes into phagocytic cells (Epstein-Barash et al. 2010; Patel 1992; Patel and Moghimi 1998), and in a mean diameter size of<200 nm, for maximizing both efficacy and safety (Allen and Hansen 1991; Epstein-Barash et al. 2010; Rodrigueza et al. 1993; Torchilin 2005).

Since γδ T cells are unique to primates (Hinz et al. 1997) our study on the effect of liposomal formulations was limited to *ex vivo* experiments. We first assessed the selective targeting of conventional liposomes to monocytes in *ex vivo* cultures of h-PBMCs, which comprise of monocytes and lymphocytes. It should be noted that the specific uptake by DC in the PBMC culture was not determined. Liposomes were selectively internalized intact by 83 ± 1.8% of h-monocytes, whereas only 6 ± 0.6% of the lymphocytes internalized liposomes following 4 h of incubation (Fig. 16.1). The scant uptake of liposomes by lymphocytes is attributed to B cells, and not T cells, as previously demonstrated for nanoparticles (Sela et al. 2010). The intracellular accumulation of IPP in monocytes, caused by nBP inhibition of FPP synthase of the mevalonate pathway, subsequently activates cytotoxic γδ T cells proliferation and secretion of pro-inflammatory cytokines (Hewitt et al. 2005). As expected from a non nBP (Kunzmann et al. 2000), treatment with CLOD-L or the free drug had no effect on γδ T cells proliferation (Fig. 16.2b). The similar stimulatory effect of liposomal and free ALN on γδ T cells proliferation (Fig. 16.2b) and TNFa secretion (Fig. 16.3) is most likely due to a similar uptake of free and ALN-L by monocytes in culture. The significantly increased cellular uptake, in tissue cultures, of particulated dosage forms of BPs in comparison to free BPs is not observed when the free BP concentration is low (<1 μM) (Monkkonen et al. 1994). This is because the cellular uptake of free BPs by monocytes/macrophages *in vitro* is mediated probably by a calcium complex formed in the culture media (Monkkonen and Heath 1993; Monkkonen et al. 1994). At low BPs concentration the calcium to BP ratio is high, enabling enhanced complexation of free ALN and internalization.

Human γδ T cells recognition of LPS is predominately presented by CD1 on APCs resulting in increased γδ T cells proliferation (Hava et al. 2005; Cui et al. 2009). ALN in liposomes and as a free drug increases the secretion of inflammatory cytokine by LPS-stimulated monocytes (Epstein-Barash et al. 2010; Makkonen et al. 1999). However, treatment by LPS reduced the stimulatory effect of ALN-L (Fig. 16.4), which is in contrast to the expected synergistic stimulation of γδ T cells by LPS and ALN co-treatment. The molecule through which IPP is presented to the γδ T cell receptor is still unknown (Clezardin 2010). It could be suggested that the presentation of LPS and phospholipids on monocytes hindered the presentation of IPP resulting in the reduced stimulatory effect of γδ T cells proliferation.

Down regulation of CD14 in monocytes (adherent cells), by both ALN as a free drug and in liposomes, correlated conversely with the proliferation of γδ T cells (Fig. 16.5), suggesting that activated γδ T cells may have induced this phenotypic change. This notion is supported by the pronounced reciprocal effect of γδ T cells stimulation on APCs (Devilder et al. 2006). Activated γδ T cells secrete cytokines such as TNF-α, which in turn down regulate CD14 expression, and cause differentiation of monocytes into DC (Eberl et al. 2009). Recently, it has been shown that amplified activation of γδ T cells can be achieved by another nBP (zolendronate) treated DC (Cabillic et al. 2010). The liposomes ability to target human imDC (Fig. 16.6 right) suggests that a positive feedback could be obtained for the amplification of γδ T cells. ALN-L will activate γδ T via monocytes, which in turn will stimulate monocytes differentiation into DC, an the latter will engulf liposomes further activating γδ T cells (Takahara et al. 2008).

We observed an anti-tumor effect of liposomal ALN in mice (Fig. 16.7), with no overt infection or side effects. A growing body of evidence suggests that the anti-tumor effect of nBP is mediated by both direct and in-direct mechanisms (Caraglia et al. 2010; Clezardin 2010; Clezardin and Massaia 2010).

Since activation of γδ T cell is excluded in mice, the anti-tumor effect in our study should be attributed to passive tumor targeting as has been demonstrated for <200 nm liposomes in a murine bearing human breast cancer (Mayer et al. 1990). A direct cytotoxic effect on tumor cells is also unlikely since no significant accumulation of non-targeted liposomal formulations is expected in tumor cells (Kirpotin et al. 2006; Shmeeda et al. 2010). Thus, it is reasonable to assume that the tumor growth inhibition by ALN-L was mediated by TAM depletion, and possibly the induced differentiation of M2 into M1 (Veltman et al. 2010). The depletion of TAM and blood monocytes by liposomal formulations of CLOD and tumor growth inhibition has been reported (Banciu et al. 2008; Hiraoka et al. 2008; Zeisberger et al. 2006). ALN-L may possess a stronger anti-tumor effect than CLOD-L since it is more potent in depleting monocyte/macrophages (Danenberg et al. 2003a; Epstein-Barash et al. 2010), and at the same time, can stimulate γδ T cells activation since it is an nBP.

In attempts to increase the circulation time of nBP, Zhang et al. synthesized lipophilic pyridinium BP, which exhibit a pronounced γδ T cells stimulatory effect following incubation with h-PBMCs *ex vivo* (Zhang et al. 2010). However, systemic side effects were noted following the activation of γδ T cells (Adami et al. 1987; Hewitt et al. 2005), and the uptake of these compounds by APCs *in vivo* has not been determined. Recently, Shmeeda et al. reported that folate-targeted liposomes containing zolendronate has a potent *in vitro* cytotoxic activity on tumor cells (Shmeeda et al. 2010). In another study it was found that zolendronate activates γδ T cells to lyse tumor cells in cell culture (Li et al. 2009). Thus, it can be assumed that treatment with a combined formulation of targeted and conventional liposomes encapsulating nBP may result in a synergistic anti-tumor effect mediated by activation of γδ T cells as well as by depletion of tumor cells and monocytes/TAM. In light of our results that ALN-L activate γδ T cells *ex vivo*, we can envisage the following picture *in vivo*: Enhanced activation of γδ T cells would be obtained due to interaction with circulating monocytes as well as by TAM endocytosing liposomal nBP, leading to a potentiated anti-tumor effect of nBP. It should be noted that this scenario is speculative since γδ T cells are unique to primates (Hinz et al. 1997). The potential anti-tumor activity of ALN-L by activating cytotoxic γδ T cells and the resultant inhibition of tumor growth, presented in this study, could be assessed by investigations in primates correlating depletion of TAM and γδ T cells activation.

References

Adami, S., Bhalla, A. K., Dorizzi, R., Montesanti, F., Rosini, S., Salvagno, G., & Locascio, V. (1987). The acute-phase response after bisphosphonate administration. *Calcified Tissue International, 41*, 326–331.

Afergan, E., Epstein, H., Dahan, R., Koroukhov, N., Rohekar, K., Danenberg, H. D., & Golomb, G. (2008). Delivery of serotonin to the brain by monocytes following phagocytosis of liposomes. *Journal of Controlled Release, 132*, 84–90.

Allen, T. M., & Hansen, C. (1991). Pharmacokinetics of stealth versus conventional liposomes: Effect of dose. *Biochimica et Biophysica Acta, 1068*, 133–141.

Banciu, M., Schiffelers, R. M., & Storm, G. (2008). Investigation into the role of tumor-associated macrophages in the antitumor activity of Doxil. *Pharmaceutical Research, 25*, 1948–1955.

Bartlett, G. R. (1959). Phosphorus assay in column chromatography. *The Journal of Biological Chemistry, 234*, 466–468.

Cabillic, F., Toutirais, O., Lavoue, V., de la Pintiere, C. T., Daniel, P., Rioux-Leclerc, N., Turlin, B., Monkkonen, H., Monkkonen, J., Boudjema, K., Catros, V., & Bouet-Toussaint, F. (2010). Aminobisphosphonate-pretreated dendritic cells trigger successful V gamma 9 V delta 2 T cell amplification for immunotherapy in advanced cancer patients. *Cancer Immunology, Immunotherapy, 59*, 1611–1619.

Caccamo, N., Meraviglia, S., Cicero, G., Gulotta, G., Moschella, F., Cordova, A., Gulotta, E., Salerno, A., & Dieli, F. (2008). Aminobisphosphonates as new weapons for gamma delta T cell-based immunotherapy of cancer. *Current Medicinal Chemistry, 15*, 1147–1153.

Caraglia, M., Marra, M., Naviglio, S., Botti, G., Addeo, R., & Abbruzzese, A. (2010). Zoledronic acid: An unending tale for an antiresorptive agent. *Expert Opinion on Pharmacotherapy, 11*, 141–154.

Clezardin, P. (2010). Bisphosphonates' antitumor activity: An unravelled side of a multifaceted drug class. *Bone, 48*, 71–79.

Clezardin, P., & Massaia, M. (2010). Nitrogen-containing bisphosphonates and cancer immunotherapy. *Current Pharmaceutical Design, 16*, 3007–3014.

Cui, Y. C., Kang, L., Cui, L. X., & He, W. (2009). Human gamma delta T cell recognition of lipid A is predominately presented by CD1b or CD1c on dendritic cells. *Biology Direct, 4*, 47.

Danenberg, H. D., Fishbein, I., Gao, J., Monkkonen, J., Reich, R., Gati, I., Moerman, E., & Golomb, G. (2002). Macrophage depletion by clodronate-containing liposomes reduces neointimal formation after balloon injury in rats and rabbits. *Circulation, 106*, 599–605.

Danenberg, H. D., Fishbein, I., Epstein, H., Waltenberger, J., Moerman, E., Monkkonen, J., Gao, J., Gathi, I., Reichi, R., & Golomb, G. (2003a). Systemic depletion of macrophages by liposomal bisphosphonates reduces neointimal formation following balloon-injury in the rat carotid artery. *Journal of Cardiovascular Pharmacology, 42*, 671–679.

Danenberg, H. D., Golomb, G., Groothuis, A., Gao, J., Epstein, H., Swaminathan, R. V., Seifert, P., & Edelman, E. R. (2003b). Liposomal alendronate inhibits systemic innate immunity and reduces in-stent neointimal hyperplasia in rabbits. *Circulation, 108*, 2798–2804.

Devilder, M. C., Maillet, S., Bouyge-Moreau, I., Donnadieu, E., Bonneville, M., & Scotet, E. (2006). Potentiation of antigen-stimulated V gamma 9 V delta 2 T cell cytokine production by immature dendritic cells (DC) and recpirocal effect on DC maturation. *The Journal of Immunology, 176*, 1386–1393.

Dieli, F., Vermijlen, D., Fulfaro, F., Caccamo, N., Meraviglia, S., Cicero, G., Roberts, A., Buccheri, S., D'Asaro, M., Gebbia, N., Salerno, A., Eberl, M., & Hayday, A. C. (2007). Targeting human gamma delta T cells with zoledronate and interleukin-2 for immunotherapy of hormone-refractory prostate cancer. *Cancer Research, 67*, 7450–7457.

Eberl, M., Roberts, G. W., Meuter, S., Williams, J. D., Topley, N., & Moser, B. (2009). A rapid crosstalk of human gammadelta T cells and monocytes drives the acute inflammation in bacterial infections. *PLoS Pathogens, 5*, e1000308.

Epstein, H., Berger, V., Levi, I., Eisenberg, G., Koroukhov, N., Gao, J., & Golomb, G. (2007). Nanosuspensions of alendronate with gallium or gadolinium attenuate neointimal hyperplasia in rats. *Journal of Controlled Release, 117*, 322–332.

Epstein, H., Gutman, D., Cohen-Sela, E., Haber, E., Elmalak, O., Koroukhov, N., Danenberg, H., & Golomb, G. (2008). Preparation of alendronate liposomes for enhanced stability and bioactivity: In vitro and in vivo characterization. *AAPSJ, 10*, 505–515.

Epstein-Barash, H., Gutman, D., Markovsky, E., Mishan-Eisenberg, G., Koroukhov, N., Szebeni, J., & Golomb, G. (2010). Physicochemical parameters affecting liposomal bisphosphonates bioactivity for restenosis therapy: Internalization, cell inhibition, activation of cytokines and complement, and mechanism of cell death. *Journal of Controlled Release, 146*, 182–195.

Ferrarini, M., Ferrero, E., Dagna, L., Poggi, A., & Zocchi, M. R. (2002). Human gamma delta T cells: A nonredundant system in the immune-surveillance against cancer. *Trends in Immunology, 23*, 14–18.

Fleisch, H. (1998). Bisphosphonates: Mechanisms of action. *Endocrine Reviews, 19*, 80–100.

Geran, R. I., Schumach, A. M., Abbott, B. J., Greenber, N. H., & Macdonal, M. M. (1972). Protocols for screening chemical agents and natural-products against animal tumors and other biological-systems. *Cancer Chemotherapy Reports. Part 3, 3*, 1–103.

Gnant, M., Mlineritsch, B., Schippinger, W., Luschin-Ebengreuth, G., Postlberger, S., Menzel, C., Jakesz, R., Seifert, M., Hubalek, M., Bjelic-Radisic, V., Samonigg, H., Tausch, C., Eidtmann, H., Steger, G., Kwasny, W., Dubsky, P., Fridrik, M., Fitzal, F., Stierer, M., Rucklinger, E., & Greil, R. (2009). Endocrine therapy plus zoledronic acid in premenopausal breast cancer. *The New England Journal of Medicine, 360*, 679–691.

Haber, E., Danenberg, H. D., Koroukhov, N., Ron-El, R., Golomb, G., & Schachter, M. (2009). Peritoneal macrophage depletion by liposomal bisphosphonate attenuates endometriosis in the rat model. *Human Reproduction, 24*, 398–407.

Haber, E., Afergan, E., Epstein, H., Gutman, D., Koroukhov, N., Ben-David, M., Schachter, M., & Golomb, G. (2010). Route of administration-dependent anti-inflammatory effect of liposomal alendronate. *Journal of Controlled Release, 148*, 226–233.

Hava, D. L., Brigl, M., van den Elzen, P., Zajonc, D. M., Wilson, I. A., & Brenner, M. B. (2005). CD1 assembly and the formation of CD1-antigen complexes. *Current Opinion in Immunology, 17*, 88–94.

Hayday, A. C. (2000). Gamma delta cells: A right time and a right place for a conserved third way of protection. *Annual Review of Immunology, 18*, 975–1026.

Hewitt, R. E., Lissina, A., Green, A. E., Slay, E. S., Price, D. A., & Sewell, A. K. (2005). The bisphosphonate acute phase response: Rapid and copious production of proinflammatory cytokines by peripheral blood gd T cells in response to aminobisphosphonates is inhibited by statins. *Clinical and Experimental Immunology, 139*, 101–111.

Hinz, T., Wesch, D., Halary, F., Marx, S., Choudhary, A., Arden, B., Janssen, O., Bonneville, M., & Kabelitz, D. (1997). Identification of the complete expressed human TCR V gamma repertoire by flow cytometry. *International Immunology, 9*, 1065–1072.

Hiraoka, K., Zenmyo, M., Watari, K., Iguchi, H., Fotovati, A., Kimura, Y. N., Hosoi, F., Shoda, T., Nagata, K., Osada, H., Ono, M., & Kuwano, M. (2008). Inhibition of bone and muscle metastases of lung cancer cells by a decrease in the number of monocytes/macrophages. *Cancer Science, 99*, 1595–1602.

Kirpotin, D. B., Drummond, D. C., Shao, Y., Shalaby, M. R., Hong, K. L., Nielsen, U. B., Marks, J. D., Benz, C. C., & Park, J. W. (2006). Antibody targeting of long-circulating lipidic nanoparticles does not increase tumor localization but does increase internalization in animal models. *Cancer Research, 66*, 6732–6740.

Kunzmann, V., Bauer, E., Feurle, J., Weissinger, F., Tony, H. P., & Wilhelm, M. (2000). Stimulation of gamma delta T cells by aminobisphosphonates and induction of antiplasma cell activity in multiple myeloma. *Blood, 96*, 384–392.

Lang, J. K. (1990). Quantitative-determination of cholesterol in liposome drug products and raw-materials by high-performance liquid-chromatography. *Journal of Chromatography, 507*, 157–163.

Lewis, C. E., & Pollard, J. W. (2006). Distinct role of macrophages in different tumor microenvironments. *Cancer Research, 66*, 605–612.

Li, J. Q., Herold, M. J., Kimmel, B., Muller, I., Rincon-Orozco, B., Kunzmann, V., & Herrmann, T. (2009). Reduced expression of the mevalonate pathway enzyme farnesyl pyrophosphate synthase unveils recognition of tumor cells by V gamma 9 V delta 2 T cells. *The Journal of Immunology, 182*, 8118–8124.

Makkonen, N., Salminen, A., Rogers, M. J., Frith, J. C., Urtti, A., Azhayeva, E., & Monkkonen, J. (1999). Contrasting effects of alendronate and clodronate on RAW 264 macrophages: The role of a bisphosphonate metabolite. *European Journal of Pharmaceutical Sciences, 8*, 109–118.

Mariani, S., Muraro, M., Pantaleoni, F., Fiore, F., Nuschak, B., Peola, S., Foglietta, M., Palumbo, A., Coscia, M., Castella, B., Bruno, B., Bertieri, R., Boano, L., Boccadoro, M., & Massaia, M. (2005). Effector gamma delta T cells and tumor cells as immune targets of zoledronic acid in multiple myeloma. *Leukemia, 19*, 664–670.

Mayer, L. D., Bally, M. B., Cullis, P. R., Wilson, S. L., & Emerman, J. T. (1990). Comparison of free and liposome encapsulated doxorubicin tumor drug uptake and antitumor efficacy in the SC115 murine mammary-tumor. *Cancer Letters, 53*, 183–190.

Miyagawa, F., Tanaka, Y., Yamashita, S., & Minato, N. (2001). Essential requirement of antigen presentation by monocyte lineage cells for the activation of primary human gamma delta T cells by aminobisphosphonate antigen. *The Journal of Immunology, 166*, 5508–5514.

Monkkonen, J., & Heath, T. D. (1993). The effects of liposome-encapsulated and free clodronate on the growth of macrophage-like cells in vitro: The role of calcium and iron. *Calcified Tissue International, 53*, 139–146.

Monkkonen, J., Taskinen, M., Auriola, S. O., & Urtti, A. (1994). Growth inhibition of macrophage-like and other cell types by liposome-encapsulated, calcium-bound, and free bisphosphonates in vitro. *Journal of Drug Targeting, 2*, 299–308.

Patel, H. M. (1992). Serum opsonins and liposomes: Their interaction and opsonophagocytosis. *Critical Reviews in Therapeutic Drug Carrier Systems, 9*, 39–90.

Patel, H. M., & Moghimi, S. M. (1998). Serum-mediated recognition of liposomes by phagocytic cells of the reticuloendothelial system – the concept of tissue specificity. *Advanced Drug Delivery Reviews, 32*, 45–60.

Richards, P. J., Williams, B. D., & Williams, A. S. (2001). Suppression of chronic streptococcal cell wall-induced arthritis in Lewis rats by liposomal clodronate. *Rheumatology, 40*, 978–987.

Rodan, G. A. (1998a). Control of bone formation and resorption: Biological and clinical perspective. *Journal of Cellular Biochemistry. Supplement, 30–31*, 55–61.

Rodan, G. A. (1998b). Mechanisms of action of bisphosphonates. *Annual Review of Pharmacology and Toxicology, 38*, 375–388.

Rodrigueza, W. V., Pritchard, P. H., & Hope, M. J. (1993). The influence of size and composition on the cholesterol mobilizing properties of liposomes in vivo. *Biochimica et Biophysica Acta, 1153*, 9–19.

Roelofs, A. J., Jauhiainen, M., Monkkonen, H., Rogers, M. J., Monkkonen, J., & Thompson, K. (2009). Peripheral blood monocytes are responsible for gamma delta T cell activation induced by zoledronic acid through accumulation of IPP/DMAPP. *British Journal of Haematology, 144*, 245–250.

Sela, E., Chorny, M., Gutman, D., Komemi, S., Koroukhov, N., & Golomb, G. (2010). Characterization of monocytes-targeted nanocarriers biodistribution in leukocytes in ex-vivo and in-vivo models. *Nano Biomedicine and Engineering, 2*, 1–10.

Shmeeda, H., Amitay, Y., Gorin, J., Tzemach, D., Mak, L., Ogorka, J., Kumar, S., Zhang, J. A., & Gabizon, A. (2010). Delivery of zoledronic acid encapsulated in folate-targeted liposome results in potent in vitro cytotoxic activity on tumor cells. *Journal of Controlled Release, 146*, 76–83.

Slegers, T., van Rooijen, N., van Rij, G., & van der Gaag, R. (2000). Delayed graft rejection in pre-vascularised corneas after subconjunctival injection of clodronate liposomes. *Current Eye Research, 20*, 322–324.

Takahara, M., Miyai, M., Tomiyama, M., Mutou, M., Nicol, A. J., & Nieda, M. (2008). Copulsing tumor antigen-pulsed dendritic cells with zoledronate efficiently enhance the expansion of tumor antigen-specific CD8+ T cells via V gamma 9 gamma delta T cell activation. *Journal of Leukocyte Biology, 83*, 742–754.

Torchilin, V. P. (2005). Recent advances with liposomes as pharmaceutical carriers. *Nature Reviews. Drug Discovery, 4*, 145–160.

van Rooijen, N., & Van Kesteren-Hendrikx, E. (2003). "In vivo" depletion of macrophages by liposome-mediated "suicide". *Methods in Enzymology, 373*, 3–16.

Veltman, J. D., Lambers, M. E. H., van Nimwegen, M., Hendriks, R. W., Hoogsteden, H. C., Hegmans, J., & Aerts, J. (2010). Zoledronic acid impairs myeloid differentiation to tumour-associated macrophages in mesothelioma. *British Journal of Cancer, 103*, 629–641.

Wilhelm, M., Kunzmann, V., Eckstein, S., Reimer, P., Weissinger, F., Ruediger, T., & Tony, H. P. (2003). Gamma delta T cells for immune therapy of patients with lymphoid malignancies. *Blood, 102*, 200–206.

Zeisberger, S. M., Odermatt, B., Marty, C., Zehnder-Fjallman, A. H., Ballmer-Hofer, K., & Schwendener, R. A. (2006). Clodronate-liposome-mediated depletion of tumour-associated macrophages: A new and highly effective antiangiogenic therapy approach. *British Journal of Cancer, 95*, 272–281.

Zhang, Y. H., Cao, R., Yin, F. L., Lin, F. Y., Wang, H., Krysiak, K., No, J. H., Mukkamala, D., Houlihan, K., Li, J. K., Morita, C. T., & Oldfield, E. (2010). Lipophilic Pyridinium Bisphosphonates: Potent gamma delta T Cell Stimulators. *Angewandte Chemie-International Edition, 49*, 1136–1138.

Zito, M. A., Koennecke, L. A., McAuliffe, M. J., McNally, B., van Rooijen, N., & Heyes, M. P. (2001). Depletion of systemic macrophages by liposome-encapsulated clodronate attenuates striatal macrophage invasion and neurodegeneration following local endotoxin infusion in gerbils. *Brain Research, 892*, 13–26.

Index

E. Zahavy et al. (eds.), *Nano-Biotechnology for Biomedical and Diagnostic Research*, Advances in Experimental Medicine and Biology 733, DOI 10.1007/978-94-007-2555-3,